AF324019

Perspectives on Water

Published in association with
Observer Research Foundation
New Delhi

Observer Research Foundation (ORF) is a public policy think-tank that aims to influence formulation of policies for building a strong and prosperous India. The ORF pursues these goals by providing informed and productive inputs, in-depth research and stimulating discussion. The Foundation is supported in its mission by a cross-section of India's leading public figures, academics and business leaders.

Perspectives on Water

Constructing Alternative Narratives

Editors

Lydia Powell and Sonali Mittra

ACADEMIC FOUNDATION

NEW DELHI

www.academicfoundation.com

First published in 2012
by

ACADEMIC FOUNDATION
4772-73 / 23 Bharat Ram Road, (23 Ansari Road),
Darya Ganj, New Delhi - 110 002 (India).
Phones : 23245001 / 02 / 03 / 04.
Fax : +91-11-23245005.
E-mail : books@academicfoundation.com
www.academicfoundation.com

in association with the
Observer Research Foundation, New Delhi

In partnership with:
Rosa Luxemburg Stiftung, Germany
(Rosa Luxemburg Foundation)
www.rosalux.de

© 2012
Copyright: Observer Research Foundation, New Delhi

Cataloging in Publication Data--DK
  Courtesy: D.K. Agencies (P) Ltd. <docinfo@dkagencies.com>

Perspectives on water : constructing alternative narratives /
  editors, Lydia Powell and Sonali Mittra.
    p.   cm.
  ISBN 13: 9788171889709
  ISBN 10: 8171889700

  1. Water resources development--Indus River Watershed.
2. Water-supply--Management. 3. Water use. 4. Water
resources development--Government policy.   I. Powell,
Lydia. II. Mittra, Sonali. III. Observer Research Foundation.

DDC 333.9100954   23

Typeset by Italics India, New Delhi.
Printed and bound in India.

Contents

Section II

Running Dry: Micro to Macro Case Studies

Section III

Water Policy Perspectives

Section IV

Sharing Water Resources

About the Editors/Contributors

Editors:

Lydia Powell, Senior Fellow & Head, Centre for Resources Management, Observer Research Foundation (ORF), works on policy issues on energy, water and climate change. She has been with the ORF Centre for Resources Management for over eight years. Powell holds three postgraduate degrees—two from the Norwegian School of Management, Oslo, on energy and one in Solid State Physics from Cochin University of Science & Technology, India. She has worked with Norsk Hydro and Orkla, two of Norway's largest listed conglomerates whose interests include oil & gas, solar energy and hydro power.

She currently edits the *ORF Energy News Monitor*, weekly newsletter on energy and routinely contributes articles on energy policy, energy pricing and regulation, climate and equity.

Over the past few years, her research interests have expanded into understanding non-traditional security aspects with particular reference to energy and water. 'Re-Imagining Indus' is among her most recent research projects, involving contributions from over 30 experts on India and Pakistan.

Sonali Mittra, Research Assistant, Centre for Resources Management, ORF, is an enthusiastic young researcher with a deep interest in policy research associated with water development and sustainability. She did her graduation from Delhi University in Botany (Hons), followed by a masters in Environmental Impact Assessment and Management from University of Manchester, United Kingdom.

With academic experience in divergent fields, she has developed an understanding of the dynamics of environmental change; appreciating the role of social responsibility in meeting the challenges of sustainable development; the institutional, regulatory and strategic basis of environmental planning and the relative merits of different environmental planning solutions. Her master's dissertation focussed on 'Sustainability in Fair Trade Agricultural Practices: Case Study India' with respect to environmental planning and management.

Over the past year, she has been working with the Resources Management team at ORF with the focus on water issues, small hydro development and solar energy in South Asia.

Contributors:

Paulette Bynoe (*paulette.bynoe@uog.edu.gy*) (BA; MPhil; PhD) is currently spending a sabbatical at the Department of Life Sciences, University of the West Indies, St Augustine Campus, Trinidad.

Dr Bynoe has 17 years of professional experience as an educator and as an environmental policy specialist. She was earlier Director of the School of Earth and Environmental Sciences of the University of Guyana (2009-2011). Currently, Member of the Regional Team of Experts for JICA/CDERA Project on Community Disaster Risk Management in five pilot countries in the Caribbean; Member of the Scientific Advisory Committee for Forestry Network Project (FORENET) for ACP countries; and an environmental consultant in community-based disaster risk management, environmental impact assessment, environmental education and training, environmental management and disaster risk reduction and climate change adaptation policy formulation at international and national levels.

Rohan D'Souza (*rohanxdsouza@gmail.com*) is Assistant Professor at the Centre for Studies in Science Policy at Jawaharlal Nehru University, New Delhi. Previously, he has been a postdoctoral fellow at the Agrarian Studies Program in Yale University, and a Ciriacy-Wantrup fellow at the University of California, Berkeley. He was also a Senior Research Associate at the Centre for World Environmental History,

University of Sussex and Visiting Fellow at the Resource Management in Asia-Pacific Program, Australian National University.

Dr D'Souza's studies on the history of hydraulic manipulation in the Indian subcontinent were published as a book titled: *Drowned and Dammed: Colonial Capitalism and Flood Control in Eastern India (1803-1946)*, and he has recently published a jointly edited volume titled: *The British Empire and the Natural World: Environmental Encounters in South Asia*. His interests and research publications cover themes in environmental history, conservation, ecological politics, sustainable development and modern technology.

Shailaja Fennel (*ss141@cam.ac.uk*) is Director of Research at Cambridge Central Asia Forum and a University Lecturer in Development Studies, and a Fellow of Jesus College at the University of Cambridge. Shailaja has been researching the linkages between rural developments, environmental and educational strategies in India, China and Central Asia since 2004. She has specialised in the sub-fields of institutional reform, rural development, gender and household dynamics, kinship and ethnicity, and educational provision.

Her recent publications include: *Rules, Rubrics and Riches: The Relationship between Legal Reform, Institutional Change and International Development* (2010); *Gender Education and Development: Conceptual Frameworks, Engagements and Agendas* (2008) edited with M. Arnot. She is currently completing a book titled: *Grains and Gains: The Political Economy of Agriculture in China and India* (2012) and working on a manuscript, currently titled: *Development in Transition: Lessons from Central Asia*.

Maaz Gardezi (*maaz.gardezi@lums.edu.pk*) is currently a Research Associate at the Development Policy Research Centre (DPRC) at Lahore University of Management Sciences. DPRC is a knowledge centre structured around core socioeconomic development themes with the objectives of carrying out cutting edge multi-disciplinary research. Maaz received his Bachelor's degree in Economics from the University of Bath, UK. Before returning to Pakistan, he worked at Porsche Cars, Great Britain. Later, he joined a health management firm in Lahore. His areas of interest in research are: environmental policy, climate change and adaptation, food security, institutional reforms and governance, and regional trade and growth.

He is a very active member of the DPRC Water Programme that focusses on a broad range of water-related issues: interprovincial water conflicts, water pricing and trading, institutional reforms and water governance, climate change and adaptation and transboundary water management.

Uwe Hoering (*uweHoering1@web.de*), living in Bonn, Germany, is a freelance journalist and policy analyst working on a wide range of development and environment issues. He lived for several years in New Delhi, India and Nairobi, Kenya, working for newspapers, journals and German radio stations. Topics he frequently covers are: agricultural development, policies for the urban and rural water sector, international financial institutions like the World Bank, and social movements, mainly in Africa, South and Southeast Asia. Currently he is focussing on foreign direct investments into agriculture and the debates and conflicts they have generated.

Among his recent publications is a book on agricultural policies in Africa (*Agrarkolonialismus in Afrika. Eine andere Landwirtschaft ist möglich*, Hamburg), a booklet on grassroots alternatives in water management (*Water to the People: Conflicts and Alternative Concepts in India*) and a collection of three case studies from Tanzania, Brazil, and Indonesia on perspectives of small scale agriculture (*Who Feeds the World? The Future is in Small Scale Agriculture*).

A.K. Jain (*akjain@pau.edu*), an agricultural engineer, is working as Head, Department of Soil & Water Engineering at Punjab Agricultural University, Ludhiana (Punjab). With professional experience of 26 years, he has authored many technical bulletins and research articles on irrigation, water management and groundwater modelling in national/international journals. He has worked as Principal Investigator of a research project on 'Groundwater Modelling and Management' funded by DST/TIFAC and as co-PI in many projects on 'water management' funded by ICAR. In 2007, he participated in a workshop on 'Sustainable Water Resource Management' at the University of Florida (USA). He was Chairman, College Academic Affairs Committee and Secretary, Board of Studies of the College. Dr Jain has received a 'Team Award' and 'Commendation Medal' from the Indian Society of Agricultural Engineers for outstanding research contributions and is also a recipient of the 'Shiksha Rattan Puraskar' for the year 2010 from IIFS, New Delhi, for outstanding contributions in education and research.

Michael Kugelman (*michael.kugelman@wilsoncenter.org*) is the South Asia associate at the Woodrow Wilson International Center for Scholars (Washington, DC) where he is responsible for research, programming and publications on South Asia. He received his MA in law and diplomacy from the Fletcher School at Tufts University, and his BA from the American University's School of International Service. Much of his work has focussed on natural resource issues in India and Pakistan. He has written and edited a number of publications on the region, including *Reaping the Dividend: Overcoming Pakistan's Demographic Challenges; India's Contemporary Security Challenges* (2011); and *Empty Bellies, Broken Dreams: Food Insecurity and the Future of Pakistan* (2011). His work has also appeared in *Foreign Policy, World Politics Review, The Times of India, Asia Times Online, Dawn, The News, Newsline, Express Tribune* and *Daily Times*.

Yu Li (*yueli@aades.org*) is a Director (South Asia) of the Asia-Africa Development Research Institute (Development Research Centre) of the State Council, People's Republic of China. He graduated from the China Institute of Contemporary International Relations in 2008 with a PhD in International Relations. He is focussing on the study of international political and economic relations of South Asia, non-traditional security issues and international security cooperation regimes.

N. Shantha Mohan (*shantha_nias@yahoo.co.in*) is currently anchor of the water programme at the National Institute of Advanced Studies (NIAS), IISc Campus, Bangalore. She is an educationist with specialisation in the area of economics of education. Over the past

25 years, Dr Mohan has been engaged in research from a gender perspective on issues related to governance, violence against women, water and literacy. Her interests also include developing new methodologies and models for intervention and policy advocacy. She has been the recipient of the Shastri Indo-Canadian Women in Development Fellowship and The National Council for Research on Women Award, USA. In the water sector, her work has focussed on issues surrounding transboundary river water sharing in India.

Sailen Routray (*sailen.routray@apu.edu.in*) teaches at the Azim Premji University in Bangalore. His research interests lie at the intersections of anthropology of the everyday state and of development with a thematic focus on the water sector. He is also interested in social sciences in Indian languages and the contemporary history of Odisha. He writes in English and Odia for both academic and non-academic journals.

Muhammad Azeem Ali Shah (*azeem.ali@lums.edu.pk*) completed his Masters at the University of Engineering & Technology in Lahore, and is currently a doctoral candidate at the Lahore University of Management Sciences (LUMS). Azeem worked with Honda Atlas Cars in Lahore, and was with Nestle Pakistan Limited, just before he commenced his PhD studies in the fall of, 2008. Azeem is also a researcher on Institutions and Policy Analysis at the International Water Management Institute (IWMI) Pakistan).

He has been actively involved in research related to water resources, reforms and flood disasters. Some of his recent projects include assisting a judicial flood inquiry tribunal in its proceedings; 'Re-imagining the Indus' , a Department for International Development (DFID) funded project on institutional and policy reforms in irrigation sector between India and Pakistan; 'Smart Water Grids' with the LUMS water group and a few collaborative research projects with the Harvard South Asia Initiative (SAI) team. He has presented his research work at local as well as international conferences.

Shawahiq Siddiqui (*shawahiq.ielo@gmail.com*) is an Environmental Lawyer and partner at Indian Environment Law Offices, a law firm based in New Delhi specialising in environmental law. Shawahiq specialises in climate change and energy law and has worked extensively on India's National Solar Mission and National Mission on Enhanced Energy Efficiency. He has been the principle investigator for drafting the Renewable Energy Law (Draft) for the Ministry of New and Renewable Energy, Government of India and the Nagaland Biodiversity Rules for the state government of Nagaland. He has also drafted the alternate Coastal Regulation Zone Notification. He has worked extensively on the Forest Rights Act, Tribal Self-Rule Law and Access to Justice for the Marginalised. He is currently working to develop an effective legal and regulatory response to the National Water Mission under the National Action Plan for Climate Change. Shawahiq is also Honarary Director at the Energy and Environment Council.

Inderjeet Singh (*inderjeetsidhu@rediffmail.com*) is a Professor of Economics at Punjabi University, Patiala (Punjab), India. He specialises in economic development and planning. His doctoral research work relates to 'Input-Output Analysis of Energy Consumption in India'. His post-doctoral work includes a large number of research papers, books and project reports. He has been associated with two prestigious research projects: 'Infrastructure Linkages between Two Punjabs' and the 'UNDP-sponsored District Human Development Report of Sangrur'. At the organisational level, he is associated with many national and international professional associations.

Shah Bakht Sohail (*s_shahbakht@hotmail.com*) is law graduate from the Lahore University of Management Sciences (LUMS), Lahore, Pakistan, with special interest in the alternative means to dispute resolution and environmental law. Apart from working at Hanif, Adnan and Faisal (HAF), a law firm in Lahore, Pakistan, she is currently assisting the instructors in teaching courses on Constitution and Administrative Law and Environmental Law and Regulation in Pakistan at LUMS. Sohail has written several research papers during the course of her studies.

Editors' Note

Water can only be defined in so many ways for it is an absolute material resource rather than an abstract notion that can lend itself to infinite interpretations. However, a multitude of definitions and interpretations drawing on its chemical, physical, economic, environmental and social attributes have evolved with the advancement in science and society. Interestingly, this hydro evolution in nomenclature has less to do with the natural order of things and more to do with human ambitions, the quest for modernity and unrestrained material development.

Desolately enough, our understanding of water and its complex manifestations remains superficial. Material development and its conflict with nature in general and water in particular is moving to the centrestage with raging debates and research papers promoting alternative narratives. Reductionist narratives that treat water challenges as simple quantifiable obstructions against the economic engine of the world, political narratives that present shallow 'ethical' juxtapositions of water as well as security discourses that dramatise the challenge to predict war among nations over water, are being constantly debated.

Through engineering feats of mammoth proportions, water resources have been harnessed to such an extent that many regions of the world have been declared to be in a state of 'water stress'. How did we manage to put the human race in such an extreme situation and who is to blame for it? At the local level we find it convenient to blame it on the politicians and their inability to plan and govern efficiently. At the national level, we find solace in blaming nature and its vagaries: inadequate monsoon is held responsible for water shortages and intense monsoon is blamed for destroying our water storage structures. In the case of water resources such as rivers that are shared across territorial borders, we find it convenient to blame the region on the other side of the border. At the global level we assign collective blame on the notion of 'scarcity' even though we do not really understand it. Scarcity is presented as the dark side of nature which wilfully places limits on our innocent pursuit of material wealth generation. As brilliantly argued by Lyla Mehta in her recent book (*The Limits to Scarcity*: 2011), the finiteness of water does not necessarily lead to its scarcity: the scarcity of water today is the result of unequal access to and control over the finite resource in a finite world. Naturalising the scarcity of water makes it 'socially acceptable' as it projects it as something that is determined by 'nature' and it makes it seem like a 'commonsensical truth'. As Mehta eloquently points out, 'scarcity of water is manufactured through historic, political, social and economic processes and institutions that have disadvantaged the poor. The poor get less but pay more and bear the brunt of human development costs too.

The 'water challenge', if we may call it that, calls for a radical change in the way we interact with nature and a change in the manner in which we define progress and our 'needs' in the context of material progress. The objective of this volume is to strike upon some of the dominant narratives from a diverse range of contributors (from around the world) to examine the water challenge through the hard lens of geology, hydrology economics and environmental science as well as the soft lens of social science, political science and international relations. The volume opens with papers on the history of the water use practices in the Indus Basin, its evolution over time and its current status and goes on to showcase some micro and macro narratives of water in general. A selection of papers on the political, social and environmental conflicts over water follows and the volume concludes with perspectives on the contentious issue of sharing water resources across territorial boundaries.

Historical perspectives on the water use practices in different river basins are inherently more informative than generally presumed. Water use practices vary with the histories and geographies of specific regions and therefore general prescriptive solutions that universalise notions, practices and experiences on water can often be irrelevant. Section I of this book traces the hydro evolution of the water-use practices in the Indus Basin from the pre-colonial to the post-colonial period. The papers highlight the transformation of the basin from technical and social perspectives. The insightful paper on hydropolitics in the Indus Basin (Rohan D'Souza;

Chapter 1) calls for re-defining and re-inventing river systems as more complex heterogeneous systems than mere homogeneous entities that can be subject to massive human interventions. The causal pathways tracing water use over 5,000 years leads to the other end of history, captured in papers on current agricultural water use patterns in the Indus Basin (Inderjeet Singh, Chapter 4 and A.K. Jain, Chapter 5). The consequence of technological interventions in river courses, unpredictable climate and unrestrained development that resulted in the floods in Pakistan in 2010 along with the reasons behind depleting water tables in Pakistan—which are strikingly different from the reasons behind excessive groundwater use in India—are examined by two authors from the country (Maaz Gardezi, Chapter 2 and Muhammad Azzem Ali Shah, Chapter 3). What is startling to read is that the agricultural revolution in the Indus, which is often held up as a model for progress, is in reality an ecological crisis in disguise. It would not be too far-fetched to say that neglecting the ecological deterioration in the region any further would be detrimental to the very survival of the Indus Basin and the civilisations that thrive around it.

Section II contains papers which are essentially case studies and micro-perspectives on how we perceive water. The message that comes out clearly is that water defines itself on the basis of the context in which it is used and it is futile to look for universalised notions. Location and context-specific evaluation adds a lot more to the understanding of the multifaceted nature of water. One of the papers in this section poses an intriguing question: 'Is water a commodity or a human right?', and proceeds to examine possible answers (Uwe Hoering, Chapter 6). Authors of region-specific case studies expose fundamental flaws in the administration and management of water provisioning in urban settings. The paper on the failure of privatisation of water supply in Europe has many lessons for mega urban centres in Asia. The case studies highlight the complexities with regard to high population density and poor clean water access in South Asia (Yu Li, Chapter 7). The papers also touch upon issues such as changing consumption patterns, industrial revolution and unstable political and administrative structures that have added to the water crisis. One surprising fact that the paper on Guyana in South America (Paulette Bynoe, Chapter 8) reveals is that even if it is at the other end of the world with an entirely different terrain, geology, culture and tradition, the water challenge the country faces is strikingly similar to that in South Asia. Optimistically, we may say that there is unity in diversity as far as water is concerned and yet pessimistically we can also say that similar patterns of influence from politically motivated and ethically juxtaposed leaders of the world produce similar sub-optimal outcomes!

More often than not, the water crisis is governed by 'absolutism' and it is this aspect that defines Section III of this volume. Political objectives influence the policies, administration and governance of water resource management and the political undercurrents define the water situation with their vested interests, ignoring both micro and macro ecological reality. Contributors to this section examine ecological sensitivity, climate change and degradation narratives that are taking root in sustainable development discourses and expose the manipulation of water issues for political objectives. The authors (Shailaja Fennel, Chapter 9 and Shah Bakht Sohail, Chapter 11) address issues such as the contestation over the common property resources of poor and vulnerable communities, constitutional arrangements for water allocation as well as efficiency in water usage and conservation. Lack of legal sanctity for the policy instruments on common property resources is highlighted (Shawahiq Siddiqui, Chapter 10) as one of the crucial factors in the ineffectiveness of constitutional arrangements. The distressing fact that the papers bring out is that common governance challenges such as inadequate monitoring and regulation and gaps in disseminating the lessons learnt from the past processes and procedures continue to pervade and plague water resource management.

Important aspects of the ecological, social and political challenges faced in water resource management and the endemic conflicts both within and between riparian countries and regions in South Asia is addressed in Section IV of this volume. What comes out clearly is that issues related to water conflicts are intricately linked to the basin-specific dynamics. Interests of the affected population and ecological integrity of the region with shared basins face major social, economic and environmental challanges (Michael Kugelman, Chapter 12 and Shanta Mohan and Sailen Routray, Chapter 13). Fortunately, none of the authors endorse the 'water wars' thesis which is gaining momentum among western think tanks and also in the security community in developing Asia. In the context of shared water resources, it is tragic to note that what is ecologically necessary is almost always politically infeasible.

Overall, the message from the volume is fairly optimistic. We do have the tools to set ourselves on an alternate path in the context of water provided our morality coincides with our political will. That being said, an alternative path will materialise only with enhanced participation of the people with a strong sense of ownership and responsibility towards the dying rivers, melting glaciers and the abuse and misuse of water.

— Lydia Powell and
Sonali Mittra

Section I

Causal Pathways: Hydro Evolution in the Indus Basin

1 | Hydropolitics, the Indus Water Treaty and Climate Change

Writing a New Script for the Indus Rivers

ROHAN D'SOUZA

Discussions on the Indus rivers have become overwhelmingly strategic. Flows have become matters of political contest, vested interests and above all else, national security. Ironically enough, such strident noises over the division of waters have mostly avoided meaningful attempts to recall the region/watershed's oftentimes troubled histories. It is as if the Indus Water Treaty (IWT) of 1960 could be almost nonchalantly deployed to snip vast flowing courses into neat divisible segments and with equal ease 'rationally' allocate immense volumes between nations. This is, a mere blunt knife approach trying to comprehensively sever and move about a complex hydrology without so much as an afterthought about disturbing delicately poised fluvial ecologies or the implications of coarsely stirring whole river-based communities.

The IWT—with this structured 'forgetting' of the Indus Basin's many pasts and varied environments—not unexpectedly, is often concluded by experts to be a 'successful' legal-technical arrangement that has suffered from frequent and exceptional political 'misperceptions'. It can, however, be more convincingly argued the other way. The IWT was an unsteady political project to begin with and is now fatally failing as a legal-technical arrangement. But reversing the analytical vantage requires a sharp perceptual shift as well. A type of taproot understanding of the IWT is urgently called for. By which, new facts, so to speak, must be dug up, sunned and differently seasoned in order to have us go beyond the limited simplifications of hydraulic data, official statistics, engineering opinion and statist imperative.

Between the 16th and the 18th centuries, the Mughal Empire held, in a single firm embrace, vast territories of what today comprises India and Pakistan. For the Mughal ruling elites, applying a regular squeeze over agricultural surpluses was the preferred route to wealth and privilege. Typically enough, given this essentially land-based notion of power, the empire's numerous and intricate network of rivers were, at best, used either for navigation or as avenues to conduct easy trade. These inestimable flows, in other words, became natural outliers to the imperial governments' otherwise more onerous quest to extract revenues from the soil. It would be unfair, however, to entirely dismiss all Mughal efforts at harnessing water. Several innovative structures, for example, helped deftly steer river currents into gardens, fountains, hunting grounds and even giant reservoirs.[1] On balance, nevertheless, comprehensive fluvial management was rarely ventured upon. It was only in the middle of the 19th century, following the steady consolidation of British colonial rule in the sub-continent, that the big immodest engineering interventions for total hydraulic control was carried out. In particular, the vast semi-arid flood plains—sandwiched between the Indus and Gangetic river systems—became amongst the first sites the world over for implementing large-scale modern irrigation schemes.

The Great Hydraulic Turn

For the sprawling Indus Basin, coursed through by the fluvial fingers of the Indus, Ravi, Chenab, Beas, Sutlej and Jhelum, colonial hydraulic interventions were, in fact, both technically and politically unprecedented.[2] For the first time in the region, permanent structures in the form of barrages and weirs were thrown across riverbeds. These durable head-works were equipped with a series of shutters to regulate flows by impounding water during lean seasons, to be then diverted in calibrated quantities across miles of canals. On the reverse, in times when the rivers were swollen or torrential, the shutters would be flipped open to hurriedly jettison discharge. In effect, by alternately impounding or quickening the

---

1. Singh (1992); Siddiqui (2008). For a useful collection on pre-colonial irrigation and water management strategies in the subcontinent, see Agrawal and Narain (1997); Wahi (1997).
2. D'Souza (2006a).

discharge of flows, the river's variable or moody regime, it was held, could be transformed from a seasonal to a perennial irrigation possibility.[3]

Beginning with the Upper Bari Doab Canal (1859) and the Sirhind system (1882), the drive climaxed with the 'most ambitious' irrigation project of the colonial period: the Triple Canal Project (1916). These perennial canal schemes, however, were assembled not merely as channels commandeering river flows but were, in the words of David Gilmartin (1994), crucially linked to 'political imperatives of state building.'[4] The colonial dispensation, in effect, vigorously pursued perennial irrigation and agricultural settlement as means essential for stabilising its otherwise unsteady authority in the region. At heart, canal building was the pressing attempt to yoke the then just disbanded Sikh soldiery and a large number of non-cultivating 'predatory' herdsmen to 'permanent interests in landed property'. The impacts of perennial irrigation, however, can also be historicised differently. Indu Agnihotri, in a seminal essay on the canal colonies in the British Punjab, argued that irrigation did not, as is widely held, simply bring water and increase agricultural productivity into hitherto desolate 'wastes'. Rather, the colonial canal colonies of the late 19th and early 20th centuries overwhelmed and overran a pre-existing vibrant pastoral economy who, besides herding, also seasonally cultivated crops through inundation canals. This process of marginalisation, if not substantial elimination, of the pastoral and their unique ways of living with the ecologies of the doabs continues to find only rare mention.[5]

The point here is that the introduction of modern irrigation in the semi-arid flood plains of the Indus system was enabled following intense struggles over the creation of landed property, the elimination of pastoral livelihoods and accompanied by relentless wide-ranging environmental transformations. Raising agricultural productivity through perennial irrigation, hence, involved, by design or otherwise, a deafening silence about different pasts: the ignored but suffered consequences of waterlogging, soil salinisation, the violence of landed property, the defeat of

---

3. Wilson (1989); Harris (1923). For an introduction to the modern hydraulic moment in British India, see Whitcombe (1972); Stone (1985); Ali (1987); D'Souza (2006b) and Hardiman (1998).

4. Also see Gilmartin (2003).

5. Agnihotri (2012).

nomadic peoples, instabilities brought on by mono-cultures and commercial agriculture, the attrition-ridden assembling of colonial social hierarchies and inevitably the forced 'training' of once volatile free-falling rivers into contained, disciplined irrigation channels.

Profoundly intertwined with the relentless march of modern irrigation in the doabs were the life and the world of the colonial civil engineer. Though often less heralded, these energetic, restless, innovative and adventurous men of the empire were made steadfast with technical training in modern river management and control. Such expertise was systematically rubbed into them at the Addiscombe Military Seminary (near Croydon), Royal Indian Engineering College (Cooper's Hill) or the Thomason College of Civil Engineering (Roorkee). Through the lens of 'imperial science', colonial environments for these engineers were not merely to be 'catalogued, studied and observed' but actively pursued for large-scale manipulation; all in the name of commerce, civilisation and endless improvement. In the same stride, this resolute quest to control nature was intimately tied to the equally severe project of dominating colonised populations. For the British colonial enterprise, in other words, intensely extracting from nature and exploiting subject people seemed almost logically to go hand in hand.

Attempting the dramatic transformation of complex and immense river systems through engineering, however, was no simple task. In aiming to physically shuffle, transfer, move or redirect vast volumes, engineers resorted to reductionist and specialised mentalities. That is, colonial engineers planned and crafted modern river control initiatives primarily through ideologies for abstractions in the form of formulas, equations, model-making, and by repeatedly fine-tuning an overwhelmingly quantitative notion of hydrology. Irrigation engineering preferred, in terms of their self-image and professional training, to be defined principally by the 'mathematical sciences'.[6] Such a notion of handling water, in effect, assumed an unequivocal trust in numbers, while simultaneously aiming to wilfully ignore and shut out local knowledges or place-specific ecological idiosyncrasies involved in harnessing flows. If anything, therefore, the ascendance of colonial hydrology meant the consolidation of the universal,

---

6. Weil (2006).

expert-driven and specialised practices for river management alongside the steady marginalisation of localised cultures and place-based knowledges for water management. Christopher V. Hill, in a recent article, notes that engineers from the Royal Indian Engineering College (Cooper's Hill) were encouraged through their training to sustain an 'ignorance of the environmental exceptionality of their respective regions, and a deep belief that local knowledge was irrelevant.'[7] The mighty Indus Basin, in effect, was disciplined with the elegance of numbers and rational hydraulic model building. The river systems, hence, that otherwise stood as messy miscible admixtures of flows, histories, cultures, localities and exceptional environments was conceptually recast as straight, contained channels. A once heterogeneous collection of people and places, through imperial science, cement and quantitative hydrology could be turned into homogeneous spaces.

Nation-Making with Cross-Border Rivers

Following the hydraulic re-arrangements of the 19th century, the Indus Basin witnessed, in the mid-20th century, a second equally dramatic rupture—the division of waters for nation-making. In effect, scuffles over hydraulic access and rights that characterised the colonial period were transformed into bitter disagreements over issues of ownership and control of the Indus waters. As the Radcliffe Line etched a hard border between India and Pakistan on 17th August 1947, flows had to be reconfigured as national rivers. From previously watering an uninterrupted contiguous political bloc, the Indus and its tributaries, in step with this logic of Partition, had to be abruptly inserted within new geographical scales and imagined as part of decolonised national biographies.

Not unexpectedly, complications over the Indus erupted as intractable hydropolitics between India and Pakistan. For a start, flows had to be instantaneously sliced and diced at multiple conceptual levels, in order to acknowledge the region's changed geopolitical realities. Stretches of the tributaries, hence, that fell within India were classified overnight as upper riparian waters while Pakistan, on the other hand, inherited the downstream flows. Having been thus officially instituted as cross-border

---

7. See Hill (2011).

flows, the various arms of the Indus system, could now only be managed through a raft of international rules and protocols. The first involved a band-aid approach, with the concluding of an immediate pact, appealingly termed the Standstill Agreement, by which all existing flow arrangements were to be maintained till 31st March, 1948. Alarmingly enough for Pakistan, the Government of India 'suspended' supplies the very next day in which the agreement officially lapsed. Though flows were eventually restored after 18 long days, the shock of being denied water not only 'seared' the Pakistani sense of entitlement to the rivers but the entire incident brutally made known to both sides that water could easily translate into severe problems of politics and power.[8] The subsequent Inter-Dominion Agreement, as a stop-gap arrangement, actually ended up further amplifying the fact that sustaining a divided fluvial system, invariably, if not urgently, needed an enduring 'final settlement'.

Following a period of staggered negotiations, the IWT was finally clinched in 1960, as a trilateral deal between Pakistan, India and the World Bank. As noted by Daanish Mustafa, the IWT process substantially mirrored the political landscape of its time. A context that was defined by extreme suspicion between the two countries, their respective location in the larger geopolitical strategies for the region and relationships that were repeatedly marred by political competition. Rehearsing elements or features of the IWT, however, would not be helpful here, as they have been competently done elsewhere. What, nevertheless, needs to be marked is the fact that the IWT was overwhelmingly a legal-technical document. A notion about flows which, on the one hand, were firmly anchored in colonial legacies for water management in the region, while, on the other, water agreements were crafted as legal protocols for nation-making. That is, flows were appropriated not on the basis of their ecological properties but rather sub-divided in order to enforce hard national borders. The Indus system, in essence, was inserted into the geopolitical calculations of a troubled region and made legible primarily as statistically tabulated hydraulic data. The physical constituency of the river regime was, thus, starkly framed simply as a network of water channels, with the aspired 'normal' defined as a seasonally determined 'average volume'. Rivers as

---

8. For an excellent discussion on the politics over the Indus Rivers between India and Pakistan, see Mustafa (2011).

national resources, hence, became facts without stories and quantities without qualities. That is, flows were not understood as organically interconnected and interacting elements of wetland ecologies, aquifers, lakes, marshes and the combined actions of innumerable tributary streams. Rather, as mere volumes contained in channels, rivers could be abstracted, diverted or interfered with to satisfy national priorities. Significantly as well, as Majed Akhter in a compelling discussion points out, the IWT sought to enforce the idea of location, whereby the three eastern arms were deemed to belong to Pakistan and the three western flows were allocated to India. A neat division that, in effect, dismissed the possibility of 'hydrological bonds' or 'hydrologic interconnectivity'.[9]

Putting Ecology into Flows

The belief that rivers are merely moving masses of water crying out to be regulated and dammed has been dramatically challenged, since the 1980s, by a fresh spirited theoretical turn amongst river ecologists. These ecologists have been convincingly able to demonstrate that fluvial regimes are complex geomorphologic, chemical and biological processes in motion. Rivers, accordingly, are made up of habitat mosaics that support a wide variety of aquatic and riparian species. And the beating heart that keeps alive the river's ecological health and viability is its natural flow regime or the flood pulse, which organises and defines the river ecosystem itself.

It is now understood that natural variable flows create and maintain a range of productive dynamics between the channel, floodplain, wetland and the estuary. The magnitude and frequency of high and low flows consequently regulate numerous ecological processes. While wetlands provide important nursery grounds for fish and export organic matter and organisms into the main channels, the scouring of the floodplains by inundations often helps rejuvenate innumerable habitats for flora and fauna within the basin. Even periods of low flow sometimes creates contexts for certain kinds of ecological benefits, by allowing certain biological organisms to thrive. A large body of recent evidence indicates that natural flow regimes of virtually all rivers are inherently variable, and that this variability is critical to ecosystem function and native biodiversity.

9. Akhter (2010).

In effect, rivers with highly altered or artificially regulated flows might in most cases lose the ability to support riverine ecological processes.[10]

By thus recasting, in fundamental ways, the manner in which fluvial processes are understood, river ecologists are now suggesting that a fresh paradigm is required for managing and interacting with such hydraulic endowments. Centrally, what is being argued, is that flows are embedded in ecological contexts and therefore transferring them through technological fixes can and often do have several unintended environmental consequences. Simple steel and concrete approaches aimed at water abstraction, diversion and interference, in other words, must give way to an entirely new spectrum of knowledges, which will treat flows as being determined by non-linear ecological qualities. Put differently, treating river as mere mute volumes is flawed both as a concept and as a water management practice.

Handling and harnessing variability and stochastic flow regimes, consequently, have become critical to shaping sustainable approaches toward river management. Equally as well, flows must be grasped as organic entities: rivers stitch together innumerable ecological processes to produce an intricate and interconnected fluvial system. The entire Indus Basin, in effect, is a collection of relationships between streams, floodplains, the head reaches, aquifers and inevitably, the chaotic delta. Small wonder then, that the so called 'success' of the IWT has resulted in the relative ecological devastation of the Indus Delta. Historically, for the Indus Basin, a rough calculation suggests that before projects for siphoning flows began in the 19th century, up to 150 million acre-feet of fresh water probably fell into the delta, along with the deposition of close to 400 million tonnes of nutrient rich fertilising silt. These immense uninterrupted volumes nourished and sustained a sprawling collection of mangroves, inlets, creeks, estuaries and other wetland ecologies.

Subsequent to projects for damming and draining of the Indus and its tributaries for agriculture, hydropower and nation-building, the amount of fresh water flowing into the delta has been steadily reduced to a lean 10

---

10. Ward and Stanford (1995); Poff *et al.* (1996); Johnson *et al.* (1995); Norris and Thomas (1999).

million acre-feet (less than 10% of historical flows). The disastrous ecological consequences of starving the delta through massive fresh water transfers have only now begun to be acknowledged. Besides eroding livelihood possibilities for close to 1.2 million inhabitants in the delta and along the coasts, the fluvial impoverishment of the delta has also resulted in several tangible negative impacts on fish breeding, damage to marine food webs, destruction of unique salt water ecological habitats, and an inestimable loss in biodiversity.[11]

By suggesting that flow variability is central to fluvial health, river ecologists have put forward a definitive challenge to the cement-steel based water-control ideologies of the contemporary civil engineer; whose entire conceptual tool kit, as pointed out earlier, was mostly drawn up in the colonial setting of the long 19th century in the subcontinent. In a similar vein, the hitherto untroubled pre-eminence of the expertise generated by giant centralised water bureaucracies such as the Central Water Commission (India) and the Indus Waters Commission (Ministry of Water and Power, Government of Pakistan) needs to, in the light of these new ecological facts, be carefully qualified and reconsidered. These institutions, with their training anchored in quantitative hydraulic data, have thus far been oriented primarily towards strategising for 'average flows'. These are technical-bureaucratic institutions, in other words, that are committed to searching for and premised entirely upon harnessing hydraulic predictabilities. Significantly enough, these centralised water bureaucracies also play crucial roles in shaping national water policies and informing political processes over the building of hydraulic infrastructure in India and Pakistan, respectively. But with variability and stochasticity as the new norm for engaging with river systems, so to speak, what becomes of these legal-technical institutions and their infrastructural technologies? Put differently, if climate change is about the intensification of hydraulic unpredictability in the region, will the IWT as a legal-technical institution be able to respond to the new challenges?

---

11. See the report by the IUCN (2003). Also see Memon (2005).

Climate Change, Changing Contexts and Changed Constituencies

Pakistan in 2010 was witness to an unprecedented weather moment. Sometime in July a 'blocking event' occurred, which technically refers to an entire jet stream being halted. In this particular case, the blocking event hovering over the Western Himalayas then ended up colliding against the then oncoming and heavily laden summer monsoon clouds. The collision led, predictably enough, to an intense precipitation episode. Immense volumes of water hurtled down the Himalayan slopes and rapidly overwhelmed and smashed through every conceivable channel of the Indus system. Such was the intensity that four months worth of rainfall fell, by one estimate, in the span of a mere few days. Parts of Northern Pakistan, in fact, even received more than three times their annual rainfall in a matter of 36 hours.[12]

Following the downpour, the oversaturated channels violently burst their banks and Pakistan's flat flood plains were, in a flash, literally roofed over with enormous sheets of seemingly unending waters. In the assessment of the geographer Kuntala Lahiri-Dutt, while a freak weather pattern and not climate change could in an immediate sense account for the Indus floods, 'whether or not, however, this blocking event was a consequence of the long-term effects of climate change rather than simply an abnormal weather pattern is impossible to answer.'[13]

Such has been the severity of the devastation brought on by the 'great floods' of 2010 that even estimates of the overall damage continue to escape a credible count. Some rough approximations by various government agencies, international organisations and relief bodies, suggest a stunning picture, to say the least. In one careful survey of the various estimates, it is indicated that 21 million people overall were affected. Close to 1,700 people or more perished and 1.8 million homes were damaged or destroyed. In its wake, the floods also rummaged through 2.3 million hectares of standing crops and brought about a loss of US $5 billion to the agriculture sector and around US $2 billion each to the physical and social

---

12. Lahiri-Dutt (2010).
13. Ibid.

infrastructure. These disturbing numbers, nevertheless, do not cover or even indicate the long-term costs for recovery and reconstruction that will be involved in meaningfully rehabilitating both the social and economic infrastructure in the region.[14] But the flood-devastated realities of Pakistan, as Daanish Mustafa and David Wrathall in a recent insightful essay argue, point to a far more striking conclusion: that the floods were aggravated and its impacts made even more ferocious because of vulnerability. Beginning with the dramatic hydraulic transformations in the colonial period, independent Pakistan persevered in creating 'a mismatch between the design assumptions of the infrastructure, such as embankments and barrages and the dynamic reality of the channels' carrying capacity.'[15] That is, Pakistan's hydraulic and social designs were geared to 'ignore the river system's natural rhythms, in return for agricultural productivity and prosperity.' Overcoming the potential dangers in such a trade-off, for them, therefore, would require a 'better tactic', which plainly stated was to 'adapt to the Indus basin's hydro-meteorological regime.'

In effect, several obvious implications can be drawn from such a perceptive and out-of-the-box assessment. For one, evidences on the ground forcefully suggests that the IWT dispensation with its attendant hydraulic infrastructure and governance architecture will be unable to either cope with or override fluvial variability brought on by the extremes of climate change events. The IWT of 1960 pursued hydraulic infrastructural development to harness 'average flows' and crafted legal mandates to manage fluvial predictabilities. The protocols for negotiations, in such an arrangement, were moreover made dependent upon a narrow band of engineering expertise and imperatives brought on by regional geopolitics.

Over recent years, as fluvial dynamism has begun to intensify and characterise river behaviour in the Indus Basin, the IWT, many feel, has already begun to reveal its wear and tear. Concerns abound, for example, over whether the felt or perceived decline in flows from the head reaches have been brought on by problems of natural variability or the more insidious possibility, according to some voices, of systematic infrastructural interferences in the upper riparian. If the hydraulic abnormal sharp seasonal

---

14. Mustafa and Wrathall (2011).

15. Ibid.: 7.

peaks or troughs in the flow regime thereby becomes the new normal, there will be obvious pressures for more agile institutional and political responses. But as we have pointed out earlier, the IWT as a legal-technical arrangement is fundamentally based on the old normal of neat averages and equally determined statistical predictabilities. Hydraulic volatility, in other words, becomes a source of disruption and dangerous disagreement rather than a point for different opportunities and renewed understandings.

Climate change and its perceived impacts, in effect, push for an active reconsideration of the IWT framework. Instead of an overt emphasis on technical and technology-based approaches, run with the narrow expertise of engineers and state negotiators, the new compact for river management/ sustainability in the region would require different social constituencies and their experiences with the Indus waters. This would involve drawing upon and fostering cooperative dialogues between riverfront communities[16] on both sides of the border; such as fisher folk, irrigation dependent farmers, river ecologists, water historians, sociologists and aquatic specialists (to name a few). These plural narratives can imbue the IWT with a much needed ecological sensitivity. By acknowledging rivers as complex ecological entities, water sharing strategies between Pakistan and India will crucially depend on how a range of otherwise ignored historical and social fluvial experiences in the region are deployed to deal with flow variability. That is, the IWT or another compelling version has to be crafted to meaningfully grasp the Indus and its temperamental tributaries as qualities of flows rather than as blocs of disconnected volumes. The current reign of cement, steel and quantitative hydrology, in other words, must urgently give way to viable dialogues over fluvial relationships and ecological process.

Water politics between Pakistan and India urgently needs an ecological dimension and must acknowledge the varied fluvial linkages, from the delta upwards. This will involve, significantly enough, harnessing the full gamut of what has been termed as confidence building measures (CBMs),

---

16. I draw upon this useful notion of a riverfront community from Sarandha Jain's wonderfully compelling book on the Yamuna river titled *In Search of Yamuna: Reflections on a River Lost* (2011). The riverfront community, she suggests, refer not only to people who live by and off the river but become a 'bridge' between land and water, river and society and as 'mediators between nature and culture'.

such as evolving soft borders, thickening track two diplomacy and actively enabling scholarly exchange. It might no longer be possible to sustain the mere division of the Indus rivers, rather we must actively seek new political and cultural possibilities for the beneficial sharing of waters.

References

Agnihotri, Indu (2012). "Ecology, Land Use and Colonization: The Canal Colonies of Punjab", in Mahesh Rangarajan and K. Sivaramakrishnan (eds.), *India's Environmental History: Colonialism, Modernity and the Nation*. Ranikhet: Permanent Black. pp.37-63.

Agrawal, Anil and Sunita Narain (eds.) (1997). *Dying Wisdom: Rise, Fall and Potential of India's Traditional Water Harvesting Systems*. New Delhi: Centre for Science and Environment.

Akhter, Majed (2010). "More on the Sharing of the Indus Waters", in *Economic and Political Weekly* XLV(17): 99-100, April 24.

Ali, Imran (1987). *The Punjab Under Imperialism, 1885-1947*. New Delhi: Oxford University Press.

D'Souza, Rohan (2006a). "Water in British India: The Making of a 'Colonial Hydrology'", in *History Compass* 4(4): 621-28, May.

————. (2006b). *Drowned and Dammed: Colonial Capitalism and Flood Control in Eastern India*. New Delhi: Oxford University Press.

Gilmartin, David (1994). "Scientific Empire and Imperial Science: Colonialism and Irrigation Technology in the Indus Basin", in *The Journal of Asian Studies* 53(4): 1132.

————. (2003). "Water and Waste: Nature, Productivity and Colonialism in the Indus Basin", *Economic and Political Weekly* 38(48): 5057-65.

Hardiman, David (1998). "Well Irrigation in Gujarat: Systems of Use, Hierarchies of Control", *Economic and Political Weekly* 33(25): 1533-44.

Harris, D.G. (1923). *Irrigation in India*. London: H. Milford, Oxford University Press. pp.5-7.

Hill, Christopher V. (2011). "Imperial Design: The Royal Engineering College and Public Works in India", in Deepak Kumar, Vinita Damodaran and Rohan D'Souza (eds.), *The British Empire and the Natural World: Environmental Encounters in South Asia*. New Delhi: Oxford University Press. p.73.

IUCN (2003). "Indus Delta, Pakistan", in *Case Studies in Wetland Valuation* 5, May.

Jain, Sarandha (2011). *In Search of Yamuna: Reflections on a River Lost*. New Delhi: Vitasta Publishing Pvt. Ltd.

Johnson, Barry L., William B. Richardson and Teresa J. Naimo (1995). "Past, Present and Future Concepts in Large River Ecology: How Rivers Function and How Human Activities Influence River Processes", in *BioScience* 45(30): 134-41.

Lahiri-Dutt, Kuntala (2010). "Indus Floods, 2010: Why did the Sindhu Break its Agreement?", September 1. See *http://asiapacific.anu.edu.au/blogs/southasiamasala/2010/09/01/indus-floods-2010-why-did-the-sindhu-break-its-agreement/*

Memon, Altaf A. (2005). "Devastation of the Indus River Delta", Proceedings, *World Water & Environmental Resources Congress 2005*. American Society of Civil Engineers, Environmental and Water Resources Institute, Anchorage, Alaska. May 14-19. pp.1-14.

Mustafa, Daanish (2011). "Critical Hydropolitics in the Indus Basin", in Terje Tvedt, Graham Chapman and Roar Hagen (eds.), *Water, Geopolitics and the New World Order* (A History of Water, Series II, Volume 3). London, New York: I.B. Taurus. pp.374-94.

Mustafa, Daanish and David Wrathall (2011). "Indus Basin Floods of 2010: Souring of a Faustian Bargain?", *Water Alternatives* 4(1): 72-85.

Norris, Richard H. and Martic C. Thomas (1999). "What is River Health?", in *Freshwater Biology* 41: 97-209.

Poff, N. LeRoy, J. David Allan, Mark B. Bain, James R. Karr, Karen L. Prestegaard, Brian D. Richter, Richard E. Sparks and Julie C. Stromber (1996). "The Natural Flow Regime: A Paradigm for River Conservation and Restoration", in *BioScience* 47/11: 769-84.

Siddiqui, Iqtidar Husain (2008). "Water Works and Irrigation System in India during the Pre-Mughal Times", in Jos Gommans and Harriet Zurndofter (eds.), *Roots and Routes of Development in China and India.* Boston: Lieden.

Singh, Abha (1992). "Irrigating Haryana: The Pre-Modern History of the Western Yamuna Canal", in Irfan Habib (ed.), *Medieval India: Researches in the History of India 1200-1750.* Delhi: Oxford University Press. pp.49-61.

Stone, Ian (1985). *Canal Irrigation in British India: Perspectives on Technological Change in a Peasant Economy.* Cambridge: Cambridge University Press.

Wahi, Tripta (1997). "Water Resources and Agricultural Landscape: Pre-Colonial Punjab", in Indu Banga (ed.), *Five Punjabi Centuries: Polity, Economy, Society and Culture, c.1500-1990* (Essays for J.S. Grewal). New Delhi: Manohar. pp.267-84.

Ward, J.V. and J.A. Stanford (1995). "Ecological Connectivity in Alluvial River Ecosystems and its Disruption by Flow Regulation", in *Regulated Rivers, Research and Management* 335/GR: 1-15.

Weil, Benjamin (2006). "The Rivers Comes: Colonial Flood Control and Knowledge Systems in the Indus Basin, 1840-1930s", in *Environment* and History 12(1): 3-29.

Whitcombe, Elizabeth (1972). *Agrarian Conditions in Northern India: The United Provinces Under British Rule, 1860-1900.* Vol. 1. Berkeley: California University Press.

Wilson, Herbert M. (1989). *Irrigation in India.* Delhi: Daya Publishing House (First published 1903). pp.78-81.

2 | A Historical Perspective of the Indus Basin

MAAZ GARDEZI

Introduction

A historical insight into the development of irrigation in the Indus Basin depicts how resources such as land, labour and water have interplayed with technologies, culture and institutions. Each era postulates a disparate meaning to water rights and water management, values and value judgments coupled with a constantly evolving set of technologies. There has been a break with the past. Previously, the ultimate goal of water resource infrastructure development was to achieve food security, political stability, profitability and economic gains. More recently, this has also incorporated other issues, such as reducing vulnerability, improving livelihoods, alleviating poverty and reducing environmental degradation.[1] This paper begins with a brief description of the development of irrigation infrastructure during the Sultans and the Mughals. Next, the colonial era is discussed along with the struggle for survival by the communal irrigation systems and traditional practices. Complex stakeholder conflicts in irrigation still prevail in modern times; where the colonial rulers are replaced by states whose approbation with multilateral donor-driven development programmes and the consequent abstinence from cultivating the existing traditional systems and roles—and not redefining institutions and the role of state and local actors.

Origin of the Indus Basin

The mighty Indus River, originating in Tibet in the Himalayas is the 22nd longest river in the world. The genesis of the agrarian civilisation[s]

---

1. Barker and Molle (2004).

along its entire length dates back to the fifth millennium BC. The origin of ploughed pastures can be traced back to 2400 BC, as depicted by the Kalibangan excavations in Western Rajasthan. It shows grids formed by channels running north-south and east-west in a pattern identical to those of the present. Then, there are the ingenious hydraulic mechanisms that were found in the bath houses of Mohenjo-Daro that should intrigue today's engineers exploring the evolution of water systems. The length of land running along the Indus River has always been more fertile than those in other parts of India. However, the land was arid except during the three-month rainy season and some form of artificial irrigation had to be developed as the population swelled. In fact, one of the reasons for the decline of the Indus Valley civilisation has been attributed to the 'character of culture that had stagnated, probably as a result of complacency impeding further efforts. This led to inadequate maintenance of irrigation channels and bunds that resulted in total system collapse.'[2]

Agriculture and Water Management: The Sultans

Agriculture was the mainstay of people's livelihood. Lands adjacent to the rivers had fertile alluvial soil that was ideal for cultivating various crops. The vastness of land and relative abundance of rainfall had made life simple and economically sound in Punjab. However, the need and requirement for irrigation varied with the climatic and physical circumstances. The necessity for perennial irrigation was inversely proportional to the amount of rainfall. Hence, the majority of this region required some form of irrigation to sustain cultivation throughout the year —or for bi-annual harvesting.

During the pre-Mughal era, the Sultans dealt with water management as an existential issue for the prosperity of the region. The first lake constructed by Sultan Iltutmish in the 12th century was a multi-purpose lake that supplied clean drinking water to the city's population as well as for growing seasonal fruits along its boundary. In the 13th century, Sultan Alauddin Khalji is credited with having taken great interest in developing lakes, tanks and cisterns for increasing farm yield.

---

2. Asrar-ul-Haq (1998).

The most important event of that time was the construction of large artificial canals during Sultan Khalji's regime. They built canals that were linked to the large rivers, which during the monsoon floods fed the canals. Of course, in those days the outcome was unpredictable and breaching of the inundated canals was a routine affair. The construction of large canals in the region between Sutlej and Delhi was pioneered by Sultan Firoz Shah Tughluq (1351-1388). He also directed his men to build a canal from Sutlej to Jhajjar, a distance of 48 cos (3 km). This was also a period of renaissance for the building of bridges, aqueducts and dams that were used to store water throughout the year. The canals assured a modicum of water supply to the arid zones of Punjab, allowing the farmers to grow two crops—the *kharif* (summer) and *rabi* (winter)—a year for the first time. Also, now they did not have to dig very deep wells. A few other large canals were built that acted as a catalyst for fast harvesting. It also resulted in the migration of employment seekers to the region.

Another important development of this period was the introduction of the Persian wheel.[3] In the 14th century, this was a luxury that only the rich and prosperous farmers could afford. Set up on the wells, the wheel's function was to push the water up directly to the newly constructed storage tanks. The wells also had its social benefits, as the income from sale of its water was given to the poor.[4] Furthermore, the Persian wheel initiated the use of metal buckets instead of earthen vessels. By the early 16th century, the wheel was adopted in almost all sugarcane and rice growing areas of Punjab.

Irrigation: Mughal Empire

Punjab, during the Mughal era, comprised five main doabs.[5] The Bai Jalandhar Doab formed the area between Sutlej and Beas, covering an area

---

3. The Persian wheel operates in deep wells. It comprises three wheels and a beam horizontally attached to a toothed wheel outside the well. To the outer end of the beam, a pair of bulls or buffalos is yoked. The animals move in a circular path, pulling the beam and thus making the wheels revolve. As the machine is turned, the buckets hanging in a chain dip, one by one, into the water. Again, they reach the top and then empty into a trough. The water flows through a drain to the fields, orchards, etc.

4. Gopal (1980).

5. A Doab is a term used for a tract of land lying between two confluent rivers.

of 50 cos (3.125 km). The second was the Bari Doab lying between the Beas and the Ravi (17 cos = 1.06 km). Third, the Rechna Doab, the most fertile region, fell between Ravi and the Chenab (30 cos = 1.87 km).[6] Fourth was Chaj Doab between the Chenab and the Jhelum River and, lastly, the Sind Sagar Doab between the Indus River and Jhelum River—a vast tract of desolate land.

Emperor Babur, the founder of the Mughal dynasty, gave in his 'Baburnama'[7] a detailed account of the prevalent modes of irrigation practices in India.[8] The Mughals constructed perennial canals that had permanent headworks. These headworks either did not extend across the entire stream or allowed the floods to pass over their crests. Perennial irrigation meant the supply of water was available throughout the year.[9] The first evidence of perennial irrigation dates back to early 17th century when a 80 km long canal was constructed by Emperor Jahangir (1605-1627) to bring water from the right bank of the Ravi to the pleasure gardens of Sheikhupura near Lahore.

From the early 16th century onwards, groundwater was extensively used for irrigating the lands. Wells and Karez (underground water channels) were dug and lifting devices such as Charas, Shaduf, Rati and Persian wheel assisted in extracting groundwater. Such innovations that increased water availability were critical for growth in an area of sparse and unreliable rainfall. The widespread use of the Persian wheel between the 13th and 19th century led to the harnessing of cattle power to wells, and hence encouraged large-scale migration and agricultural settlement in Punjab. The system of canal inundation freed the farmers from dependence on the vagaries of river floods and allowed them to follow a regular pattern of irrigation.

To understand what went into the construction of canals in the pre-British era, one needs to link it with the political imperatives of the monarchic reign. The expansion of inundation canals, which require relatively little in terms of permanent headwork, was not always a matter of

---

6. Akbar (1985).

7. Fahlbusch *et al.* (2004).

8. Khan *et al.* (2006).

9. Ahmad (2000).

technical innovation.[10] With the decline of the Mughal Empire, the rulers turned widely to canal construction to consolidate their local supremacy and create an agricultural base through which to control the local and regional elite. This largely depended on the ability of the rulers to rally the support of the elite and their followers to get the canal work going. The emphasis was on creating communities which engaged in the sharing and maintenance of the canal and its water.[11] In the past, the prosperity of the cultivator depended solely on the benevolence of the king. The Mughals had shown they were keen to shed such autocratic attitudes by making revenue collection and assessment unbiased and by providing a momentum to agricultural production through the development of irrigation works.

Community irrigation systems—prevalent in the Indus Basin—have often been complimented for their successful balancing of social integration and traditional wisdom. With the British canal colonisation, these systems now faced risk of being exposed to new dangers—world trade, a move from subsistence to commercial farming and the subsequent competition for water.

British Canal Colonies

Soon after the annexation of Punjab in 1849, the British saw tremendous opportunity in canal construction as a catalyst for surplus food production, land revenue generation and exports. As a consequence of famine (especially in 1877), British had to revise their objectives for irrigation in Punjab from purely monetary benefits to include food security and famine prevention. There was now a trade-off between revenue generation on the one hand, and famine prevention and social stability on the other. The British distinguished their productive works from protective works.[12] By 1921, 'productive works' developed on 7 million hectares had yielded a net revenue of 9 per cent against 1 per cent for 0.3 million hectares of 'protective works.'[13] Debates in those times held diverging views

---

10. Gilmartin (1994).

11. Nijjar (1968).

12. Stone (1984).

13. Protective irrigation systems are based on scarcity by design, spreading the water thinly over a large area, regardless of the degree of scarcity experienced.

on the objectives for irrigation infrastructure construction and management and control. While some favoured production and revenue as the main objective, others argued that protective measures should serve as the panacea for sustainability (both financial and social). The debates between the British government, the Government of India, local government, revenue officers and canal engineers were hotly contested, as irrigation had become an extremely lucrative investment and the control over water resources constituted political and economic dominance.

The coming of the British to power in some ways reshaped the patterns of canal construction. The canal colonies formed by the British provided momentum to growth that lasted for more than a century. These colonies highlight the linkages between canal building, agricultural settlement and political control. The British government had failed to provide any relief for the crop failures in 1824-25 and 1832-33. Moreover, half a century later, during the famine of 1877, no real provisions were made for mitigating the effects of the natural disaster. Whatever safety nets that existed had been set up by the village community from the knowledge they had gained from their ancestors. However, due to the failures of the price mechanism—price control and taxes—and crop production, the rulers failed in their task to give any relief to the cultivators. More importantly, the wisdom inherent in traditional common knowledge was shelved in favour of modern thinking.

The irrigation network post-1885 was based on perennial canals that led off from headworks. It is estimated that between 1885 and 1947, there was an increase in the area of land irrigated by canals from 3,000,000 to 14,000,000 acres in Punjab (excluding princely states). After the annexation of Punjab, the first major canal project was the Upper Bari Doab Canal (UBDC). This was to be built at the centre of the old Sikh majority areas of Amritsar, Lahore and Gurdaspur. It provided employment to the disbanded Sikh soldiers. Besides, giving them the incentive to work towards agrarian development, it reduced the threat of an uprising in the newly annexed state.[14] The British also wanted to assert their authority and stabilise the 'frontier regions'. For example, in the Upper Sindh frontier, the new canals worked as a catalyst for the hostile tribes to busy themselves in

---

14. Mustafa (2001).

agricultural occupations. The British policy also fostered investment in canal development by big landlords and pastoral chiefs—a sound measure to ensure the maintenance of stability and peace in the region.

The colonialists restructured the canals in two stages. In the first phase, they renovated the old inundation canals and in the second, these were gradually converted into perennial canals. The Punjab inundation canals were Khanwah, Katora and Grey canals on the Sutlej River. Between 1876 and 1885, 13 other canals were constructed that drew water from the Sutlej. The Sidhani canal on the Ravi was reconstructed to irrigate the land around Multan district. In Sindh, Mitharao was the first canal built along the Eastern Nara channel of the Indus. The weir on the Jamrao canal (perennial) across the Nara River irrigated the Tharparkar district. Ghar canal was another important canal, which originated 50 km downstream of Sukkur and followed a natural offshoot of the Indus. The Sukker canal branched off from the right bank of the Indus and followed an old channel.

Large construction projects were undertaken in the Indus Basin that now forms part of Pakistan. The Lower Chenab canal was the first canal constructed by the British to bring irrigation water to lands resided by the pastoral nomads.[15] The ambitious plan to turn the Sandal bar area (now Sheikhupura, Toba Tek Singh and Faisalabad) in the Rechna Doab into a large, organised colony tested the engineering skill and dexterity of the British. A weir was erected across the Chenab at Khanki village in Gujranwala district to control the water supply to the main canal. With a release of 311 cusecs, it was one of the major irrigation projects in 1892. The Lower Jhelum canal project was sanctioned in 1897 to bring the waters from a weir at Rasul to irrigate the area between the Chenab and the Jhelum (Chaj Doab). The canal opened in 1901 and by the next year, the harvest rose by 20 per cent.

The Triple Canal Project, which was endorsed in 1905, was the first project to carry water from one river to another. This project curtailed the transfer of available water in the Jhelum River across two doabs. After the First World War, the Sukkur Barrage Project, the first barrage on the Indus, was built and was commissioned in 1932. During 1921, the Sutlej Valley Project was launched to boost irrigation in the region comprising Punjab,

---

15. Arthur (1967).

Bikaner (now in India) and Bahawalpur. The project consisted of four weirs on the Sutlej River at Ferozepur, Sulemanki, Islam and Panjnad; 11 more canals were built by 1933.

The Trimmu Barrage was the last barrage built prior to the Second World War. Located below the junction of the Jhelum and the Chenab Rivers, it was to be completed in 1937. Under construction at the time of Independence were the Kalabagh Barrage (Jinnah), Kotri Barrage on the Indus River on the Pakistan side and, on the Indian side, the Bhakra Dam on the Sutlej.

The Supremacy of the State

The British administration held revenue generation from agriculture to be an integral part of the Indian economy. At the turn of the 20th century, about 1/6th of the total revenue of Punjab was from agriculture.[16] The British government considered it its prerogative to have a share of the annual produce of each unit of land. Under the Mughal regime, the precise share of the state varied according to local conditions—weather, variability in soil fertility, average price variations in the cropping pattern and irrigation potential. Furthermore, the method of assessment and revenue collection was undertaken by different ranks of the ruler's employees and dependents. Finding this extremely flexible system useful, the British embraced some of its aspects.[17]

The British aimed at developing a capitalist-oriented colonial market. There was a drastic break with the past, evidenced by the introduction of new intermediaries between the state and the tax-paying masses. These men were usually induced to abandon traditional restraints, and hence to discard the conventional recipes that helped mitigate the burden on the typical village peasant. Collection of taxes in cash and not in kind and the half-hearted attempts to divert tax income towards the maintenance of capital greatly undermined the positive effect of the system based on traditional knowledge that had till then helped build and preserve water-management techniques in the region. The Permanent Settlement process failed in its aim to create a capitalist landowner class. Furthermore,

---

16. Habib (2005).

17. Banerjee (2005).

transfer of power from *zamindars* to these new agricultural capitalist intermediaries led to the diminishing of the peasant's occupational rights and his customary right to fixed revenue rates.[18]

In the Mughal era, the tax burden for irrigation was based on a balanced, land-water combination system. When the British decided on a policy of taxing the land without balancing it with water availability, several problems arose. One significant outcome was a decline of the traditional irrigation system in colonial India. The introduction of the new revenue generation mechanisms made the till now clear roles and responsibilities of tenants and landlords ambiguous.[19] The shift from taxing variable agricultural production to fixed yields and then the shift from tax in kind to tax in cash 'upset the rhythm of procedures' among the agrarian stakeholders. Also, with no clearly defined allocation of roles, the servicing and maintenance of tanks and channels was neglected. The above moves to create a capitalist mechanism of high profits and revenue, commercialisation of produce and diminishment of community control over water and land resources led to serious downgrading of the rightful status of the native population and caused serious harm to the operational health of the irrigation system.

Tradition *versus* Modernity

The British were, however, not oblivious to the fact that traditions played a pivotal role in Punjab's agricultural practices. In 1891, the Revenue Secretary of the Punjab government wrote to the centre:[20] 'It seemed essential to preserve the tradition of Punjab as a country of peasants [and] farmers. No other general frame of society is at present either possible or desirable in the Province..., as already remarked, capitalist farming in general is not a system suitable to Punjab. But a moderate infusion of the capitalist element is not out of advantages. It supplies natural leaders for the new society; it gives opportunity to the Government to reward its well deserving servants, and to encourage the more enterprising of the Provincial gentry; it attracts strong men who are

18. Baden (1907).
19. D'Souza (2006).
20. Habib (2005).

able to command the services of considerable bodies of tenants; it furnishes a basis from which agricultural improvements may be hereafter extended, and lastly, it enables Government to obtain a better price than might be otherwise possible for the ownership as distinct from the users of land.'

The British rulers considered perennial canal construction as a fine example of ingenuity in engineering technology. Technicians, urban mercantilists and intermediaries believed that science had given a huge boost, both to the crop yield and the land area for farming. Nevertheless, there was a negative side to perennial canals: it increased waterlogging and salinisation. With irrigation systems, much like most other large-scale technological development, a complete picture of its pros and cons become visible only with the passage of time. Perennial canals did not require annual, large-scale recruitment of labour for silt clearance, as was necessary for inundation canals. Unpaid *chher* labour[21] provided by the irrigators during the winter clearance season were considered by the British as a form of coercive slavery that undermined the guarantees of private property. However, this form of labour was a symbol of community cohesion that had been practiced for many generations. J.B. Lyall (who became the Punjab Lieutenant-Governor) was one of the few Englishmen who supported the system of *chher* labour: '...not only on grounds of economy of management, but also on the ground that it tends to preserve and promote self-government. The system is solidly founded on custom, and suits the habits and circumstances of the people concerned.'[22] On the whole, constant technical monitoring, administration and modelling had replaced the regular employment of labour. In the case of silt control, labour that was previously arranged by the community was now replaced by a more complicated technique in accordance to the evolving environment.

A perspective on the development of irrigation during the colonial era depicts a case where public irrigation systems had evolved faster than the institutions needed to manage them. There is also a realisation that the sense of community ownership, traditional communal wisdom and cooperative behaviour in irrigation could not evolve to stay abreast with sophisticated irrigation design and construction. Historically, the colonial

---

21. Whitcombe (1972).

22. Lyall (1882).

rulers were able to bifurcate community managed systems from the state-run systems. After Partition in 1947, the national government and international donor agencies, instead of strengthening participatory irrigation management, simply ignored the existence of well-functioning communal modes of irrigation and the rich traditional knowledge that it brought to the command areas of the Indus Basin.

Box 2.1

Water Distribution: Warabandi and Chakbandi

The passage of the Canal and Drainage Act (1873) was the result of the engineering community's concern over the efficiency of the existing water management system. The Act prescribed the number of outlets from each channel and the allocation of irrigated area based on each of the outlets. Such drastic centralisation and control of water management was bound to critically affect the role of the community in the design and use of irrigation systems. The British policymakers were aware of the need to sustain the balance of power between the land and its users (Ali, 1988). They were also keen to bridge the gap between science and the existing distribution systems prevailing in India. One such system of water distribution, known as *warabandi*, was based on the notion of fixing of the timed turns for water. Systematic turns based on time had always been part of the well water sharing process. The crucial difference—one which interested the British engineers the most—was that the *warabandi* system was not dependent on ties of kinship or genealogy. It treated water sharing as a single and controlled entity. For the engineers, this meant that each outlet of the channel could now be quantified as per timely use based on the farmer's need. More importantly, it meant that scientific knowledge and quantifiable data could now seamlessly be integrated with an existing system that enforced local cooperation and harmony.

Stemming from this Act, major reforms were brought forward that affected not only the new perennial canals, but also the old inundation canals. One such reform focussed on the assimilation and modification of channels and on reduction in the number of watercourses and outlets. This process was known as *chakbandi*. This system was originally introduced to tackle silting of channels, but it provided the British with a subtle advantage in having greater control over the distribution of water. But then there arose a paradox (Ali, 1988). The desire to bridge traditions and scientific modernity was itself a deterrent for the colonialist ruler's political and economic goals. Their overemphasis on approbation of the traditional and social institutions came in the way of bringing to bear on the agriculturist the practice of the capitalist system. A similar viewpoint argues that traditional communal knowledge formed the foundation of the political system and that it was a necessary prerequisite for scientific engagement in canal construction. However, political troubles (especially in the context of colonial irrigation) arose from the differences in the ideology and scientific authorities of the state itself. What was not a reason for conflict was the contradiction between engineering and the cultural imperatives of India's social organisation.

In modern times, increasing water scarcity brings forth greater attention on how water resources are properly managed. The publicly managed surface irrigation systems are already dealing with issues of sustainability as budget constraints, and inflexible institutional arrangements, are making existing irrigation projects obsolete. Furthermore, conflicts exist between stakeholders—the state, multilateral donor agencies on the one hand and local water managers and farmers on the other. Each side, although they share similar objectives, hold diverging viewpoints on how to achieve those goals.

Way Forward

Although it may seem important to discuss the development of irrigation in the post-Independence era, it has been left purposely to understand how this historical perspective can guide policymakers for redefining roles and responsibilities of institutions, local stakeholders and the state. Perry (2004) argues that there are five critical elements of successful management: first, clearly and widely available knowledge and information of resource availability. Second, policies that govern water resource development and assign priorities for users and uses. Third, transforming those policies into clear allocation rules such that users and sectors are prepared for any hydrological condition. Fourth, establishing institutional roles and responsibilities to effectively deliver irrigation services. And finally, the infrastructure required for distributing water services to each user and sector.

As explained earlier, the primary goal of irrigation development in the colonial era was political domination, famine protection, revenue and exports. In the new era of globalisation, new issues have transposed such objectives, like aiming to attain food security, livelihoods, and protection of environment, global markets and exports. The availability of resources is now a crucial issue. The growing scarcity of water has been followed by a decline in profitability for cereal grain production, technological advances as a result of national budget constraints have shrunk and there are considerable socioeconomic and demographic changes in the rural economy. We now face a very different geophysical and socioeconomic environment, where sustained food security and environmental protection is at the forefront of policy debate. The five elements constitute the pillars, which should be at the vanguard of how we respond to growing biophysical, socioeconomic and geopolitical challenges.

From a geopolitical perspective, the allocation of water between nation states continues to be the biggest challenge for Pakistan and India. The Partition of 1947 split the jurisdiction over the Indus plains between two nations, India and Pakistan. Law and politics, coupled with vaguely defined allocation rules (prior to the 1960 Indus Waters Treaty) was a bone of contention for both sides. It needs to be noted that the Lower Indus region possessed the single largest system of irrigation works in the world. Reports prepared by the Indian Irrigation Commission between 1901 and 1903 regarded the Punjab projects within the context of the entire Lower Indus Basin. The outcome of these reports was that the Indus River and its many tributaries came to be regarded as a single regional hydrologic system, which could transfer water across hundreds of miles from areas of abundance to those suffering acute shortages. As the great single-purpose system expanded, political disputes over the use of its water arose among the concerned states and even among districts. These disputes were the heritage of Partition for Pakistan and India, but what is far more to the point is that what had been developed by the British to function as a single unit, the gigantic irrigation project system, found itself split between two nations. For 13 years after Independence, the two countries adhered to the system with the hope of ultimately finding a durable solution, despite glitches and the hostile attitude towards each other. Both the countries condemned the Radcliffe Award in Punjab, for, while it gave the authority over the headworks in some districts to India, it allocated the entire network of the basin's canal originating from the headworks to Pakistan. Prior to Partition, the British had given top priority to the equitable distribution of the Basin waters. But it had not laid down the rationale for the rightful management of waters of the upper and lower riparian.

From a national perspective, in present time, the significant challenge of water management is to make compatible two paradoxical trends. First is the centralising trend for an effective, integrated management of the Basin. This requires forming a water commission or regulatory organisation at the top. On the other hand, there is a need for devolution or decentralisation, which requires management at the lowest level. With decentralisation, it becomes important to revitalise community-based management, which encourages the optimum use of local knowledge for sustainable use of water. It can be postulated that many of the present pressing concerns of the water sector, such as *abiana* collection (cost recovery), sustainability, water rights etc., will continue to be neglected due to a top-down process of governance, which are mostly driven by external development partners. What would certainly be more desirable is the adequate evolution of the respective roles of states, markets, communities and civil society. A match between institutions and physical, biological and cultural environments is possible only when the people concerned are able to fully participate in the process of institution building.

It is widely accepted that for effective institutional building clearly defined water entitlements or rights are essential. However, in the Indus Basin, with a large number of small landholders, the subject becomes extremely complex and has not yet been executed with the same success as in the developed world. Rather, water allocation is strongly influenced by political intervention and bureaucratic processes that use a top-down mechanism for assimilating and manipulating water rights. The task for policymakers is to redefine the role of state and local actors so that no right is damaged and the balance between efficiency and equity is addressed by transparent negotiations and at all levels of management.

References

Ahmad, Shahid (2000). *Indigenous Water Harvesting Systems in Pakistan.* Water Resources Research Insitute.

Akbar, M. (1985). *Punjab under the Mughal Raj.* Vanguard Books.

Ali, I. (1988). *The Punjab under Imperialism, 1885-1947.* Oxford University Press

Arthur, Aloys (1967). *The Indus Rivers: A Study of the Effects of Partition.* New Haven, Conn.: Yale University Press.

Baden, B.H. (1907). *A Short Account of the Land Revenue and its Administration in British India: With a Sketch of the Land Tenures.* Oxford: The Clarendon Press.

Banerjee, A. (2005). "History, Institutions, and Economic Performance: The Legacy of Colonial Land Tenure Systems in India", *The American Economic Review* 95(4): 1190-1213, September.

Barker, R. and F. Molle (2004). "Evolution of Irrigation in South and Southeast Asia", *Comprehensive Assessment Research Report* 5. Colombo, Sri Lanka: Comprehensive Assessment Secretariat.

D'Souza, Rohan (2006). *Water in British India: The Making of a 'Colonial Hydrology'.* Cambridge: Cambridge University Press.

Fahlbusch, H., Bart Schultz and C.D. Thatte (2004). *History of Irrigation, Drainage and Flood Management.* New Delhi: International Commission on Irrigation and Drainage.

Gilmartin, David (1994). "Scientific Empire and Imperial Science: Colonialism and Irrigation Technology in the Indus Basin", *The Journal of Asian Studies* 53(4): 1127-49.

Gopal, Lallanji (1980). *Aspects of History of Agriculture in Ancient India.* Varanasi. pp.116-120.

Habib, Zaigham (2005). "Pakistan: Indus Basin and Water Issues", *South Asia Journal* (Pakistan) 6.

Haq, Asrar-ul (1998). *Case Study of the Punjab Irrigation Department.* International Irrigation Management Institute Lahore, Pakistan National Program.

Jurriëns, Rien, Peter Mollinga and Philippus Wester (1996). "Scarcity by Design: Protective Irrigation in India and Pakistan", *Working Paper.* Wageningen: Wageningen Agricultural University.

Khan, S., R. Tariq and C. Yuanlai (2006). "Can Irrigation be Sustainable?", *Agricultural Water Management* 80: 87-99.

Lyall, James B.(1882). *Memorandum by Financial Commissioner, Punjab.* August 2. Punjab Revenue and Agriculture, Irrigation. London.

Mustafa, D. (2001). "Colonial Law, Contemporary Water Issues in Pakistan", *Political Geography* 20: 817.

Nijjar, B.S. (1968). *Punjab under the Great Mughals (1526-1707 AD).* Bombay: Thacker & Co.

Perry, C.J. (2004). "Non-State Actors and Water Resources Development: An Economic Perspective", *Non-State Actors and International Law* 3(1): 99-110.

Siddiqui, Iqtidar (1986). "Water Works and Irrigation System in India during Pre-Mughal Times", *Journal of the Economic and Social History of the Orient* 29(1): 52-77.

Stone, Ian (1984). *Canal Irrigation in British India: Perspectives on Technological Change in a Peasant Economy.* Cambridge: Cambridge University Press.

Talbot, Ian A. (2000). "The Punjab under Colonialism: Order and Transformation in British India", *Journal of Punjab Studies* 14(1): 3-10, Spring.

Whitcombe, E. (1972). *Agrarian Conditions in Northern India: The United Provinces Under British Rule, 1860-1900.* University of California Press.

Wilson, H.M. (1989). *Irrigation in India.* South Asia Books.

3 | The Effects of Modernity and Development on Water Mangement

Lessons Learnt from 2010
Floods in Pakistan

MUHAMMAD AZEEM ALI SHAH

There is no doubt that climate change played a powerful role in the 2010 floods that displaced more than 14 million people in Pakistan, causing a humanitarian crisis considered to be bigger than the combined effects of the three worst natural disasters to strike the country in the past decade. The culprit, the great Indus River, can also be considered a victim of the weather phenomena, for it can hold just so much water. Floods are nothing new to this river, which has given its name to one of the earliest civilisations that we know of—the Indus Valley Civilisation. There is nothing to compare with the sheer abundance of its flow during monsoons. But what occurred in August 2010 had never been witnessed before. More than half of the normal three-month monsoon rains fell in only one week, resulting in a flow which was in excess of the normal level by several times.

It is now an accepted fact that climate change is largely a result of rampant consumerism, in the pursuit of which the human race has relegated nature to the status of being merely a provider of resources for human consumption. This capitalist thinking is a result of its need/greed to generate ever increasing profits to sustain economic growth. So, it can be plausibly argued that the blame for the alarming change in weather patterns all over the world can be put squarely on this blind, relentless drive for economic growth.

Pakistan's 2010 floods is inarguably the most salient exemplar—a warning that cannot be ignored—of what we can expect if we are to continue with this unsustainable consumerism-development syndrome.

The devastation caused by the floods also tell a lot about the type of development path that Pakistan has followed over the years, as also its sociopolitical structure and the massive inequalities in resource distribution.

Contrary to the claims of Islamabad—or of the late Richard Holbrooke for that matter, who described this as an 'equal opportunity disaster'—the devastation was neither equal nor inevitable. There is no direct correlation between the intensity of the floods and the destruction it unleashed. Indeed, what is unbreakably tied to the unprecedented destruction is the country's unequal social structure and the developmental process.

Take, for instance, the intensity of the floods. In northwest of Pakistan, where the flood originated, devastating scenes of destruction were witnessed. Bridges, houses and other man-made structures were swept away like paper boats by the sheer force of the torrents. This intensity was in a great part a result of the deforestation that has, and is, balding many regions of Pakistan. River water flows much faster when there are no trees on its way to hinder it. Similarly, since the trees' roots are no longer holding the top soil down, the chances of landslides multiply. The villain of the piece in this deforestation is the 'timber mafia' (the supplier for the other businesses). No less guilty are the government officials who, all for a quick buck, allow the mafia's men to get away with rampant pillaging of the forests, one of the country's richest resources.

Similarly, the fact that water levels stayed at more than eight feet height for several days in places like Nowshera has much less to do with the quantity of water flow than with the fact that all kinds of modern construction had mushroomed up, trapping the water in what was once a flood plain. The stagnant water was the cause for the spread of diseases like cholera, among others.

The state's laissez faire policy has allowed the unhampered exploitation and abuse of nature in various ways, causing great harm to the environment (pollution). One of the offshoots of the government's rather arrogant thinking is the attempt to 'modernise' the 'primitive' communities, in which these innocent people are made a part of the activities (such as felling trees) with scant regard for the resultant ecological disaster. That deforestation could multiply the negative impact of floods does not appear to be a major concern of the Environment Protection Agency (EPA), a body propped up to showcase the government's 'aim' to balance development with environmental health. It is no surprise that the Agency has achieved very little results even though it has been around for 13 years and has many 'qualified' people on board. All it has done is to launch standard environmental awareness campaigns, or implement the National Environmental Quality Standards (which has been superseded by ISO standards). Connecting forests or flood plains to communities' livelihoods appears to be beyond its mandate.

In fact, the entire government machinery is configured in such a way that it makes coordination of work on the larger goals far more complex and difficult. Forests come under the Forest Department, and environment under the EPA. Floods are the concern of the Meteorological Department (Met), the Irrigation Department and the newly created Disaster Management Authorities. Different barrages are operated by different entities. Coordination among these organisations is only on paper, rendering their existence a mere ritual of governance. Take the Met, for instance. Just prior to the flood, the different divisions of the Met, i.e., Flood Forecasting Division (FFD), National Weather Forecasting Centre (NWFC) and Research and Development (R&D), each came up with varied forecasts confusing everyone about what the weather would be like. The NWFC started issuing forecasts from mid-July that there were some unusual weather conditions developing in the Bay of Bengal. FFD was silent on what the consequences could be till as late as 27th of July 2010. It was only when they had real-time data that they issued their first qualitative forecast, but even this did not mention the possibility of floods. As for the R&D Department, it had not issued even a single annual report about its research on climate change, global warming and the possible vagaries of the coming monsoon. It confined itself to the publication of the usual academic journal.

What happened at the barrages presents further evidence of the bureaucratic nature of state institutions. Barrages were the most vital man-made structures that the torrents encountered on their way to the Arabian Sea. Barrages raise the water level in rivers so that irrigation canals can be fed and, in the process, the natural course of the river is diverted. Because of the pressure the river would exert on the sides when its course is being changed, they are first strengthened by training works, guide bunds and marginal bunds. These structures must withstand the water pressure if nearby dwellers are to be protected. Because of their critical importance, standard operating procedure dictates that all maintenance and development works on these structures must end before the flood season starts on 15th of June every year. This basic rule was broken in the case of Jinnah Barrage located at Kalabagh. An emergency repair work started over a year ago on the barrage's downstream apron had not been completed when the floods hit the barrage in late July. Because of the work in progress, about 10 gates out of a total of 56 gates of the barrage were closed, leaving 46 gates to take all the pressure. This resulted in increased pressure on the barrage and its allied structures as the closed gates obstructed the flow of water. When the Irrigation Department's men tried to get these gates opened, powerful waves of water came swirling to flow along the side the Left Guide Bund, which collapsed under the pressure of the barrage, leaving the nearby villages to bear the first onslaught of the floods.

Next, the torrent hit further downstream, on the recently rehabilitated Taunsa barrage. With World Bank funding, this barrage had recently been equipped with a state-of-the-art control system to operate the gates. But, incredibly, the control room was not operational at this critical moment. Foreign agencies had met the entire expense of Rs 600 million and had also installed the control system but it proved to be too 'high tech' for the department's untrained technicians. The whole blundering affair is a typical example of hurrying up the modernisation process without carrying the concerned department and its staff along—thereby leaving them helpless and useless in the moment of crisis.

The National Disaster Management Authority (NDMA) (with its provincial branches), created following the 2005 earthquake in Kashmir, is no less responsible for this state of unpreparedness. Political rivalry between the Centre (dominated by the Pakistan People's Party) and the Punjab government (ruled by the Pakistan Muslim League—Nawaz) meant that Punjab, Pakistan's most densely populated province, was the only one without a disaster management authority. At the Centre too, things had stopped moving after the Authority was set up. It is almost five years since it was commissioned, but neither a national-level plan nor a district-level disaster management authority have been formulated or created.

The issue of being prepared aside, the political dynamics during the floods clearly revealed the bias inherent in the governance of the state. As the mighty Indus flows through Pakistan, it irrigates vast swathes of land in Punjab and Sindh and therefore influences the political shape of things in this traditionally agrarian country. And, as can be expected in any capitalistic democracy, most of the land on both sides of the river is owned by prominent political figures of the country, who have vested interests in protecting their area of cultivated land from any kind of eventualities/calamities. These interests came to the fore as the Indus threatened to flood areas several miles to its left or right. State resources were blatantly used to build bunds and breach canals. The result was always the same: exposing the areas and people most vulnerable to floods while protecting those who were the least threatened. While the practice was universally applied, the poor people of Sindh, perhaps Pakistan's most feudal province with negligible land reform structures, suffered the most. Inhabitants of cities such as Khairpur, Karampur or Jampur paid with their lives to save the properties of their feudal masters.

All of the state's resources were effectively put at the disposal of the landed elite. If the poor wanted to save themselves or access these resources, they could only do it through the feudal lords in their districts. The system in Pakistan, at the best of times, is based on political patronage. During the floods, this became the sole succour for the poor—reinforcing their subordination to the landed elite.

At the state level too, this patronage syndrome was highlighted during the catastrophe. The government blatantly exploited the situation to extract more loans and grants out of countries that the state was in the habit of accusing for not respecting its sovereignty. The relief and rehabilitation work that ensued with the pledged $1.7 billion and the millions that were collected in charity was completely uncoordinated. Months after the flood, the affected people were still living in camps waiting for the government to help them. The Centre used this opportunity to enforce new inflationary taxes—at long last fulfilling a promise it had made to the IMF long ago. The post-flood debate in the country's political set-up has largely focussed on enforcing flood taxes and the reform of general sales taxes. It left the people wondering: if the reformed GST was such an important issue, then why did the government wait for the floods to implement it? Naomi Klein (2007) called it 'disaster capitalism' and she wasn't far off the mark. The politicians have used the relief funds in a blatantly partisan manner to oblige their constituencies. Corporate honchos have claimed brownie points for showing a sense of social responsibility. The army has used it to redeem their much tarnished image and the Americans to further their campaign to win 'hearts and minds' campaign. All this was water off the duck's back for the common man.

In sum, the retreat of the Indus floods, leaves behind a gloomy picture—the desperate plight of the common people, their chronic dependence on their feudal masters—on the societal canvas, of inequality and oppression. Nothing has really changed in Pakistan. The flood will soon be forgotten and, with it, the displaced families who, having lost their livestock and abodes, are still struggling to survive one day to the next. It is doubtful that the state will discard its modernising craze, fuelled as it is with borrowed money, to shift to a community-based, eco-friendly model of development. It is also unlikely that when the next flood comes—as it certainly will— there will be any less misery. In fact, the way things are going, the vulnerability of people will only have increased by then. Even if the Met is able to get its act together and the Irrigation Department manages to fortify all the barrages, unmitigated poverty will still ensure that the millions inhabiting the river banks and living in shacks are entirely at the mercy of the floods and feudal lords. The forests, meanwhile, will continue to disappear, and unplanned construction will continue to litter the flood plains, preparing traps for flood waters. And, whichever government that will be in power, it will still be waiting to burden the hapless public with new taxes. However disastrous the floods may be, it is still less calamitous than the on-the-ground results of the inequitable and oppressive policies of the Pakistan state.

Reference

Klein, Naomi (2007). *The Shock Doctrine: The Rise of Disaster Capitalism.* London: Allen Lane. pp.558.

4

Ecological Implications of Green Revolution in Punjab

With Special Reference to Water Resources

INDERJEET SINGH

Introduction

Modern economists emphasise the catalytic role that technological changes play in the growth of an economy. These changes bring about an increase in per capita income, either by reducing the amount of inputs per unit of output or by yielding more output for a given amount of input (Mathur, 1953). Technological change in an economy, therefore, refers to changes in the input-output relations of production activities. Consequently, as the economy moves from lower to higher stages of development, there occurs a shift from simpler to more modern and complicated techniques of production on the one hand and ecological fallouts on the other (Leontief, 1963). International evidence indicates that, though agriculture revolution has ensured much-needed food security, it has, concomitantly, raised alarm signals on the ecological front.

Worldwide, the groundwater quantity is decreasing and the quality degenerating as a result of the agricultural revolution (WHO, 1990). Next in harm's way is the health and survival of all living creatures. From an ecological perspective, heavy and indiscriminate use of chemicals and pesticides has contaminated the surface and groundwater, damaged fisheries, destroyed freshwater ecosystems, entered the food chain in a subtle way and the very existence of mankind is facing extreme danger (Dung and Dung, 1999; Dung *et al.*, 1999; Huan and Thiet, 2000; Kishi *et al.*, 1995 and Tardiff, 1992). Punjab, the food grains storehouse of India spearheaded the green revolution. Although, the state's abundance in agriculture output has led to higher per capita income and better standards of living, the price it has had to pay has been the increasing fall in the water table and groundwater overdraft. This paper is an attempt to analyse the ecological fallout of the agriculture model followed by the state, with special reference to water.

Analysis

The state of Punjab has an area of 50,362 sq km falling between latitude 29°32'–32°28' and longitude 73°50'–77°00'. Presently, there are 20 districts and 141 blocks in the state. It is the most developed state of India where all the villages are electrified and connected by metalled roads. The terrain is flat alluvial plain, except for the narrow mountainous belt along the northeastern border and, in the southwest, sand dunes dotting the landscape. The climate is semi-humid to semi-arid. The rainfall decreases progressively from 125 cm in the northeast (Dhar Kalan) to about 30 cm in the southwest (Ferozepur). The monsoon rains come between July to September followed by a long dry spell that puts pressure on man-made irrigation systems. The groundwater level varies from almost near surface to about 65 m below ground level. The deep water levels are in the Kandi belt and waterlogging conditions exist in some parts of the southwestern districts.

Punjab is a part of the Indus River system in the north and north-west of the Indian subcontinent. It is separated from the Ganga Basin by the Ghaggar River. It flows only seasonally and is infamous for its flash floods in the southeastern parts of the state. Other significant perennial rivers of the Indus system that flows through Punjab are the Ravi, Beas and the Sutlej that together carry 40.5 x 10^9 m^3 of water. Himalayan glaciers melt account for about 58 per cent of the source of these river waters. All these rivers are tapped by dams at different levels in the catchment areas and the stored water is utilised for irrigation through a strong network of canals in the command areas. The canal system provides water to the neighbouring states of Haryana and Rajasthan.

Problem

The advent of the green revolution in the mid-60s brought about a radical structural change in Punjab's agricultural system, with a strong chemical orientation. Traditional agriculture had progressively given way to modern and commercial agriculture. The output of wheat and rice had increased manifold, with the focus on increasing production, especially that of food grains. Apart from high yielding varieties of wheat and rice, what facilitated the process was the consolidation of landholdings, expansion of irrigation facilities, higher use of chemical fertilisers and pesticides, farm mechanisation, power and road infrastructure and easy access to inputs, as also a market support mechanism for the output. The adoption of this strategy has raised many development-related problems on economic, social and environmental fronts.

Punjab is an agrarian state with 85 per cent of its geographical area under cultivation and an average crop intensity of 189 per cent. Water is the only natural resource available and the state has no other mineral or natural resources. Intensive farming has led to a heavy demand for water and modern cropping patterns has led to immense strain on the irrigation system. All the surface water resources are being fully utilised through well-organised canal irrigation systems, but it falls far short of the farmers' needs. They therefore turn to groundwater, causing a great strain on the resource which, it must be remembered, is a basic need not just of farmers but all consumers i.e., the industry, power plants and households. Thus the state is facing a dual phenomenon of rising and falling water tables. In the southwestern region, it is rising because water extraction is limited due to the blackish/saline quality and it is falling in northwestern, central, southern and southeastern areas because the groundwater is generally fresh and fit for irrigation. This has far-reaching implications for the ecology of the region.

Availability and Deficit of Surface Water

Punjab, the major riparian state, has a limited share of water in its three perennial rivers (Sutlej, Ravi and Beas). It has been allocated only 1.795 million hectare metre (14.54 MAF)[2] out of a total average availability of 4.24 million hectare metre (34.34 MAF)[2]. Its replenishable groundwater resources are estimated at about 2.144 million hectare metre (17.37 MAF)[2]. The total available water resources are 31.91 MAF against an estimated demand of 50 MAF,[3] showing a deficit of 38 per cent for a major riparian state. Continuous growth in population, sowing of high water-consuming and high-yielding cash crops, as also the expansion of economic activities has blown up the demand for water out of all proportion to its availability.

Punjab has a total of about 14,500 km-long canal network (Table 4.1) and about 1 lakh km of watercourses, providing irrigation to 1.15 million hectare, which is 28.19 per cent of the total cultivable area of the state (Year 2006-07P). However, the network of canals, which is more than 150 years old, is unable to take its full discharge, as it stands in need of major repairs. As a result of its reduced carrying capacity and the decreasing availability of surface water, the net area irrigated by canals has gone down from 55 per cent in 1960-61 to 28 per cent in 2006-07.

At present, the canal water allowance, which has been in vogue since long, is 5.5 cusec per thousand acres in the Eastern Canal system and 3.5 cusec per thousand acres in the Sirhind Feeder system. (Both are getting water- logged.) But it is 1.95 cusec per thousand acres in the Bist Doab Canal system, which is facing depletion of groundwater. The canal water allowances need to be diverted from waterlogged areas to areas facing depletion.

Table 4.1

River Water System in Punjab

Headwork	River	Canals
Nangal headwork	Sutlej	Bhakra Main Line Canal Anandpur Sahib Hydel Channel
Ropar headwork	Sutlej	Sirhind Canal Bist Doab Canal
Shah Nehar canal system	Beas	Mukerian Hydel Channel Kandi Canal
Madhopur headwork	Ravi	UBDC Canal Kashmir Canal
Harike headwork	Sutlej and Beas	Rajasthan Feeder Sirhind Feeder
Hussainiwala headwork	Sutlej and Beas	Bikaner Canal Eastern Canal

Source: Government of Punjab. *Statistical Abstract* (various issues).

Agriculture in Punjab is primarily based on artificial irrigation, which involves using surface as well as groundwater resources. Intensive agriculture, based on wheat-rice rotation, has led to a serious imbalance in use and availability of ground resources. The total water supply of 3.13 m ham falls short by 1.27 m ham of the total water demand of 4.40 m ham (Table 4.2). The deficit is met by overexploitation of groundwater reserves through tube wells and wells.

Table 4.2

Status of Water Resources in Punjab

Detail	m ham
Annual canal water at headworks	14.54
Annual canal water at outlets	1.45
Annual groundwater available	1.68
Total annual available water resources	3.13
Annual water demand	4.40
Annual water deficit	1.27

Source: Jain and Kumar (2007).

As a result, groundwater has become a major source of irrigation in the state. To relieve the stress on groundwater, a greater emphasis is needed on an efficient conveyance and distribution system for optimal utilisation of available surface water. Punjab also needs to be given greater share in its river waters to decrease stress on groundwater resources and power consumption.

Water Use by Source of Irrigation

An analysis of net area covered by the two sources of irrigation (Table 4.3) is indicative of the fact that only 28 per cent of the total area is irrigated by surface water or canals and that the rest, 72 per cent, is by tube wells and wells. The historical dependence on canals and other sources of surface water has gradually been reduced in favour of groundwater. In Punjab, there are only two major sources of irrigation; the government canals and tube wells and wells.

Prior to the green revolution, there was an equal dependence on both the sources of irrigation (Table 4.3). The net area irrigated by canals came down from 44.53 in the year 1970-71 to 42.28 per cent in 1980-81. It slightly rose to 42.47 per cent in the year 1990-91. It has settled down to around 27 to 28 per cent in the last few years. On the other hand, because of easy availability of cheap or free electricity, the dependence on groundwater has drastically increased, especially during the decade of the

1990s. Presently, as much as over 70.68 per cent of the net area irrigated in Punjab is dependent on tube wells and wells, i.e., the groundwater and, as pointed out earlier, it is being overexploited to meet increasing demand from industry and power generation and, of course, for drinking.

Table 4.3

Net Area Irrigated in Punjab by Source

('000 Hectare)

Year	Govt. Canals	Private Canals	Tube Wells	Others	Total
1970-71	1286 (44.53)	6 (0.21)	1591 (55.09)	5 (0.17)	2888 (100)
1980-81	1430 (42.28)	- -	1939 (57.33)	13 (0.38)	3382 (100)
1990-91	1660 (42.47)	9 (0.23)	2233 (57.12)	7 (0.18)	3909 (100)
2000-01	1002 (24.92)	- -	3017 (75.03)	2 (0.05)	4021 (100)
2002-03	1148 (28.45)	-	2880 (71.38)	7 (0.17)	4035 (100)
2003-04	1129 (28.03)	- -	2889 (71.72)	10 (0.25)	4028 (100)
2004-05	1101 (27.29)	7 (0.17)	2919 (72.34)	8 (0.20)	4035 (100)
2005-06	1134 (27.93)	4 (0.10)	2914 (71.77)	8 (0.20)	4060 (100)
2006-07	1148 (28.19)	- -	2878 (70.68)	46 (1.13)	4072 (100)

Note: Figures in parentheses denote the percentages.

Source: Government of Punjab. *Statistical Abstract* (various issues).

Adding to the complexity of the problem of the tube wells proliferation (Table 4.4), there are two types of tube wells—diesel and electric—the latter being much more in demand. The total number of tube wells has risen from: 1.92 lakh in 1970-71; 6 lakh in 1980-81; 8 lakh in 1990-91; 9.3 lakh in 2000-01 and 12.76 lakh in the year 2008-09. So, over a span of four decades, the number of tube wells in use has grown by more than six times. A break-up of demands for the two types of tube wells shows that while the number of tube wells that run on diesel has remained fairly stable, the number of electric ones has increased by almost 10 times in the last 30 years and its share has crossed the 80 per cent mark. The greater part of this increase happened in the current decade. Towards the end of the 1990s, the state announced a huge concessional policy leading to the provision of free electricity to the farmers.

Table 4.4

Number of Tube Wells in Punjab

(Lakh)

Year	Diesel Operated		Electricity Operated		Total
	No.	*Per cent*	*No.*	*Per cent*	
1970-71	1.01	52.60	0.91	47.40	1.92
1980-81	3.20	53.33	2.80	46.67	6.00
1990-91	2.00	25.00	6.00	75.00	8.00
1998-99	1.70	88.54	7.45	81.42	1.92
1999-2000	1.70	18.38	7.55	81.62	9.25
2000-01	1.70	18.18	7.65	81.82	9.35
2001-02	1.75	18.42	7.75	81.58	9.50
2002-03	2.91	25.30	8.59	74.70	11.50
2003-04	2.88	25.17	8.56	74.83	11.44
2004-05	2.88	24.66	8.80	75.34	11.68
2005-06	2.88	24.14	9.05	75.86	11.93
2006-07	2.80	22.73	9.52	77.27	12.32
2007-08	2.75	22.07	9.71	77.93	12.46
2008-09	2.80	21.94	9.96	78.06	12.76

Source: Government of Punjab (2009). *Statistical Abstract.*

This subsidy further increased the use of this labour-free equipment, putting an unprecedented strain on groundwater.

Groundwater Draft and Fluctuations in Water Table

There is a dire need for a systematic, rational policy to regulate the use of this long overexploited resource. As per Table 4.5, out of the 137 blocks of the state 103 blocks are 'overexploited', 5 blocks are 'critical', 4 blocks are 'semi-critical' and 25 blocks are in the 'safe' category. A look on the temporal dimension of blocks categorisation shows that in 1984 only 44.92 per cent of the blocks were 'overexploited' and about 49 per cent were 'semi-critical' or 'safe'. By the year 1992, 53 per cent of the blocks went into the category of 'overexploitation' and the share of 'semi-critical' and 'safe' went down to 40 per cent. Presently, as per the 2004 statistics, the ratio of 'overexploited' blocks has gone up to 75.18 per cent and that of 'semi-critical' and 'safe' blocks has shrunk to 21 per cent.

Table 4.5

Categorisation of Blocks on the Basis of Groundwater Draft in Punjab

Category of Block	1984		1986		1989		1992		1999		2004	
	No.	%	No.	%	No.	%	No.	%	No.	%	No.	%
Dark (overexploited)	53	44.92	55	46.61	62	52.54	63	53.39	73	52.90	103	75.18
Dark/critical	07	05.93	09	07.63	07	05.93	07	05.93	11	07.97	05	03.65
Grey/semi-critical	22	18.64	18	15.25	20	16.95	15	12.71	16	11.59	04	02.92
White/safe	36	30.51	36	30.51	29	24.58	33	27.97	38	27.54	25	18.25
Total	118		118		118		118		138		137	

Source: Central Groundwater Board, Punjab.

The long-term water table situation (pre-monsoon) is depicted in Table 4.6. The long-term water table fluctuations for the period 1979-1998 indicated decline in nearly 76 per cent area of the state. During this period, the fall in water table was 0-3 m in 35 per cent of the area, between 3-5 m in 17 per cent of the area and more than 5 m in 24 per cent area of the state. Between years 1984-2002, there was a decline of water level in 78 per cent area of the state. During this period, water table fall was 0-3 m in 34 per cent area, between 3-5 m in 21 per cent area and more than 5 m in 23 per cent area. The maximum fall was 12.92 m in Shaina block of Sangrur district, followed by 10.85 m in Nakodar block in Jalandhar district. The maximum depth of the water level is 69.92 m in Mahilpur of Hoshiarpur, followed by 38.18 m in Hoshiarpur-II block.

As per 2004 estimates of the Central Ground Water Board, the depth of the groundwater level in the state was 0-10 m in an area of about 53 per cent, 10-15 m in approximately 30 per cent; above 15 m in around 14 per cent. The rest of the area, 3 per cent, falls in the hilly category. The lowest depths were reached in Amritsar, Jalandhar, Ludhiana, Patiala, Fatehgarh Sahib, Sangrur and Moga districts. A continuous fall of groundwater levels is occurring in about 94 per cent of the blocks.

Table 4.6

Temporal Water Table Fluctuation in Punjab, India

Year	Range of Fall (% of State Share)		
	0-3 m	*3-5 m*	*>5 m*
1979-1980	49	30	03
1979-1994	34	23	29
1979-1997	39	18	20
1979-1998	35	17	24
1979-1999	31	21	20
1984-2000	43	22	07
1984-2001	39	24	16
1984-2002	34	21	23

Source: Central Ground Water Board, Punjab.

The disaggregated picture of the block level scenario of groundwater exploitation is given in Table 4.7. The state's overall ground level development is 145 per cent. As earlier noted, out of 137 blocks, 25 fall in the 'safe' category, out of which 9 blocks are safe because the water is too saline to be of any use, such as in Bhatinda, Faridkot, Ferozepur, Mansa, Moga and Muktsar. In Bhatinda and Mansa, more than 68 per cent area has poor quality water. The proportion of such areas in Mansa district is 68 per cent, 42 per cent in Muktsar and 30 per cent in Faridkot.

Rainfall and Recharge of the System

The rainfall (positively) and rice area (negatively) determine the recharge quantum during the monsoon season. The average recharge (Table 4.8) during 1974 to 2005 was around 1 m in the rice zone in Majha and Doaba, but little less than 0.5 m in Malwa. The average recharge in Majha and Malwa remained about the same during 1990-2005 as it was during 1974-1987, but improved significantly in Doaba where it doubled during 1990-2005, when compared to 1974-1987. This rise was primarily due to investment in watershed programmes in Kandi area and the floods of 1988. The average *rabi* withdrawal has also changed significantly over time. In the Majha zone, it was 1.03 m during 1974-1987 which was intensively irrigated even earlier. It increased to 1.18 m during 1990-2005, i.e., by about 15 per cent. But the increase was about 100 per cent in Malwa, from 0.44 m to 0.85 m, and 140 per cent in Doaba, from 0.67 m to 1.60 m. The Doaba region not only has the lowest canal irrigation (2.4 per cent as compared to 39.3 per cent and 29.7 per cent in Malwa) but is also now known for its highly water-intensive crops in *rabi* season, such as sugarcane, potato, sunflower and, lately, winter maize. The first three crops covered 16.1 per cent of irrigated area in Doaba region, as compared to 4.5 per cent in Majha and only 1.9 per cent in Malwa.

It is high time that the water-use efficiency was improved by the use of several alternatives, such as appropriate planting time, irrigation scheduling, mulching, tillage, weed control and land levelling.

Table 4.7

Categorisation of Blocks according to Groundwater Draft in Punjab, 2004

District	No. of Blocks	Overexploited	Critical	Semi-Critical	Safe	State of Groundwater Development	Total Groundwater Assessment (Hectare)	Poor Quality Area (Hectare)	Poor Quality as % of Total Area
Amritsar	16	16				152	498670		
Bhatinda	7	4			3	93	354720	241200	68.00
Faridkot	2	2				106	141860	43500	30.66
Fatehgarh Sahib	5	5				161	111670		
Ferozepur	10	7	1		2	105	544190	63300	11.63
Gurdaspur	14	7	2	1	4	107	351310		
Hoshiarpur	10	2		2	6	85	333140		
Jalandhar	10	10				254	263350		
Kapurthala	5	5				204	161810		
Ludhiana	11	10	1			144	358690		
Mansa	5	5				175	207090	140900	68.04
Moga	5	5				62	217220	109100	50.23
Muktsar	4				4	178	265610	112700	42.43
Nawanshahr	5	3			2	175	132540		
Patiala	9	8		1		165	378260		
Ropar	7	2	1		4	93	207950		
Sangrur	12	12				183	508900	85000	16.70
Total	137	103	5	4	25	145	5036980	795700	15.80

Source: Central Ground Water Board, Punjab.

Table 4.8

Rainfall, Recharge, Rabi Withdrawal and Change in Water Table, Punjab

Year/Period	Rainfall (mms)	Average Recharge in Monsoon (Metres)			Average Change in Water Table (Metres)			Rabi Withdrawal (Metres)		
		Majha	Doaba	Malwa	Majha	Doaba	Malwa	Majha	Doaba	Malwa
1988	1123	2.25	3.58	1.86	0.40	1.40	0.92	1.86	2.17	0.94
1990	755	1.71	1.99	1.37	0.40	0.24	0.46	1.31	1.74	0.91
1995	794	1.77	2.96	1.38	0.46	1.22	0.55	1.31	1.74	0.84
1997	709	1.35	2.46	0.73	0.14	0.66	-0.14	1.21	1.80	0.87
1999-2005	430	0.53	0.67	-0.10	-0.61	-0.83	-0.87	1.14	1.50	0.77
Avg. 1974-1987	645	0.96	0.58	0.38	-0.07	-0.09	-0.05	1.03	0.67	0.44
Avg. 1990-2005	539	0.89	1.21	0.42	-0.29	-0.39	-0.43	1.18	1.60	0.85
Avg. 1974-2005	603	0.95	0.99	0.44	-0.19	-0.23	-0.26	1.14	1.22	0.70

Source: Singh (2007).

Demand Side of the System

On the whole, the ground area underlain with unusable water is around 7,957 sq km, which comes out to be 16 per cent of the state's territory. In addition, the state has gone from growing a previously healthy mix of crops such as wheat, maize, pulses and vegetables to devoting nearly 80 per cent of its crop area to rice and wheat, two of the most water-intensive crops.

Nation-wise, the Central and state agriculture policies—with the focus on minimum support prices, effective procurement of selected crops, input subsidies benefitting farmers in electricity, fertiliser and irrigation, as also the increased availability of credit facilities over the years—has played a key role in pushing farmers to grow primarily wheat and rice, at enormous detriment to water resource sustainability in the country.

The supply from water resources having been irreparably limited, the only hope lies in thrift, which means efficiently managing the demand and use of water. Rice has been the most remunerative crop among the *kharif* crops. As shown in Table 4.9, it is also the most water-intensive crop, using about 24,000 cubic metres of water per hectare, which is about six times more than maize, almost 20 times more than groundnut and about 10 times more than pulses.

Table 4.9

Water Requirements of Different Crops in Punjab, India

Crop	Water Requirements (Cub m per ha)	Electric Motor (Hrs per ha)
Paddy	24181	290
Wheat	5504	60
Maize	5474	50
Barley	4486	35
Kharif pulses	2355	35
Gram	2243	30
Rabi pulses	2187	30
Groundnut	1123	35

Source: Singh and Jain (2002).

The blame for the sinking water table goes to the state government's long-standing policy of giving free power to farmers. As power in Punjab is heavily subsidised, its 11 lakh agricultural consumers feel free to run their powerful submersible motors to draw groundwater, directly leading to the scarcity of this vital resource. The farmers are not concerned that the water level is already at a record low. They continue to dig deeper installing deep submersible pumps with heavy-duty motors, consuming more and more power. In the politics of water, everyone seems blind to the ground realities. A few more years of this freebie policy will render large terrains in Punjab and elsewhere barren, with not even water for drinking, leave aside irrigation.

Punjab, unfairly, is expected to be the provider of grains to the nation. Since that is the case, the Centre should compensate the state in the same manner as states with mineral reserves are given royalty on coal and bauxite. There is a point to this argument; Punjab's virtual water exports amount to 20.9 billion cubic metres every year. 'Virtual water' refers to the water embedded in commodities. For instance, a kg of *basmati* rice takes up to 7,500 litres of water to produce. An equivalent amount of water, therefore, can be deemed to have been 'exported' along with the rice. Food-surplus states are usually water exporters and the food-deficient ones, like Bihar, the importers. Keeping in view the alarming situation of overexploitation of groundwater as described above, there is an urgent need to rewrite the guidelines on the granting of power connections for agriculture pumpsets.

Rice production has been the greatest beneficiary of the minimum support price and cheap power, which the farmers got for free between 1997 and 2002. Out of the 11.44 lakh units, 8.56 lakh are electric tube wells. But all this generosity has done more harm than good. Studies have estimated that even if power is priced at production cost and charged on use basis (metered readings), rice would still be a very profitable alternative (Gill, 2003). To restore the water balance, Punjab needs to grow the 'less thirsty' crops, such as groundnut, maize, *arhar* and *moong*, particularly in the Malwa region. Besides saving water and power, it would check soil degradation, increase its fertility and bring about better sanitation and environmental conditions in the long run. A big chunk of the state budget is used up on buying costly power from other states or is diverted from high value-added sectors to this endeavour. Such spending and the ways to slash it down needs to be studied.

Floods and Waterlogging

Despite the infrastructure of dams and large headworks on all major rivers and low dams on over-discharging rivulets, the seasonal flood overloads cannot entirely be diverted and collected upstream of the dams; they have to be passed downstream so as not to endanger the dams. Sometimes the excess water has to be used for power generation even when it is not needed for irrigation. Table 4.10 shows the damage caused by such floods in the immediate past. Almost every year, several towns and villages are hit by floods, causing loss of life, cattle and land.

Table 4.10

Affect of Floods during Rainy Season in Punjab in India

Year	Villages/Towns Effected (No.)	Area Affected (in sq km)	Population Affected (No.)	Human Lives Lost (No.)	Cattle Heads Lost (No.)
1980	1191	489	85724	44	117
1990	755	471	90465	13	275
2000	81	127	319	5	88
2003	43	47	25	3	0
2004	480	610	60157	15	511
2005	93	31	125	11	48
2006	442	211	405933	10	23
2007	1033	1035	405911	7	3
2008	2001	5004	389116	34	104

Source: Government of Punjab. *Statistical Abstract* (various issues).

Also, the non-perennial River Ghaggar is prone to flash floods during monsoons. Table 4.11 gives the distribution of flood-prone villages based on the extent of damage. Extent of damage may be defined as the area damaged as a percentage of total area of a region. With an overall damage of 60.50 per cent, Andana block is the worst hit in Sangrur district. The figure stands at 66.36 per cent for Moonak (urban) and at 59.53 per cent (rural). The affected rural area constitutes 25 villages. Uniformly, the extent of damage is more than 40 per cent. The table reveals that, year after year, the *kharif* season floods have been hitting the same villages.

Block-specific plan with short, medium and long-term perspective is essential for flood control and water management, equally important being the annual repairs to and construction of structures to tame the floods. In these tasks, special care has to be taken of Ravi and Sutlej, because they run along the border and the constructions carried out across the border (as also the shifting course of the rivers) may pose a threat to our side of the river. The two nations need to coordinate their approach to the problem. What is also needed is a well thought out annual strategy for the maintenance of drains, as also to ensure the optimum utilisation of the created hydropower and irrigation potential.

About 1.04 lakh hectare area out of 2.16 lakh hectare area of Muktsar has been rendered useless because of waterlogging and, in low-lying areas, the land lies submerged. As a result, agricultural production in the area has declined over the last few years and the quality of the soil has badly deteriorated.

Water Pollution

Another challenge is the quality of water, which has been made dangerously impure by flow into the *nallahs* and rivers of untreated industrial effluents and sewage. The problem is further compounded by the mixing of storm water and sewage in various municipal towns, as these carry solid waste, biomedical waste and other hazardous wastes from city roads into the water bodies. The pollution and contamination of water resources due to industrial waste, sewage and excessive use of chemical/pesticides in agriculture has led to high pH, BoD, DO, faecal coliform and concentrations of arsenic etc. At some places, it has become toxic due to high concentration of heavy metals, such as, mercury, copper, chromium, lead, iron, nickel, cyanides and pesticides like DOT, BHC, endosulfan and Aldrin. This can cause cancer, skin diseases and miscarriages, etc. Toxic water may even enter the food chain and alter genotoxicity or damage the DNA, causing irreparable loss to all forms of life. Groundwater too is polluted due to the natural release of selenium and fluorides in some places. Improvements in the existing strategies and innovation of cost-effective techniques based on modern technology are needed to eliminate the pollution of both surface and groundwater resources. Technology and training are the key words for maintenance and improvement of this vital resource.

Health Effects

Free electricity and easy water availability has acted as a catalyst for enhancing the use of chemicals. Quantum of chemical inputs like fertilisers and pesticides are the most precise determinant of the extent of ecological damage. Chemical fertiliser consumption in Punjab is obviously high, as is seen from the fact that, with only 1.57 per cent share of the nation's total geographical area, the state is consuming 15 per cent of the country's pesticides and more than 8 per cent of chemical fertilisers. The diseases and ill-health are rising at an alarming rate and the blame falls on the indiscriminate use of chemicals. The statistics which have emerged from the recorded sales figures are bad enough, but what is worse is the ground reality of pesticide retailing in the state—a 'free for all' affair. The worst hit is the low literacy, low income populace of the cotton growing region. It has helped the pesticide traders make the area a dumping ground for chemicals. The trade of a plethora of unrecorded and spurious products flourishes here during the peak season. Pesticides (especially persistent organic pollutants or POPs) that were banned a decade ago are freely available and being actively used. Farmers and retailers are concocting their own cocktails, quite unaware of the ecological fallout.

Table 4.12 gives the data of fertiliser and pesticide-related ailments and the per capita net state domestic product. An interesting fact that comes to light is that the states that excelled during the green revolution are also at the top of the list of consumption of fertilisers and pesticides consumers, with Punjab leading all the other states. In ailments too, next to Kerala, rural Punjab leads with 136 persons per thousand, as compared to 53 in Bihar, 57 in Rajasthan and 77 in Orissa. The urban health situation is relatively better at 107 per thousand. So, the total health ailments incidence in Punjab is almost 2.5 times of what it is in Bihar. It is followed by West Bengal, Uttar Pradesh and then Haryana. Chemicalisation of agriculture has no doubt raised the per capita income, but it is certainly not worth the cost to health it has caused.

The presence of POPs in blood samples and cancer concentrations found (Thakur *et al.*, 2008; CSE, 2005) in the cotton belt is a clear indication of the ill-effects of pesticides. The ICMR report on cancer cases registered in major medical institutions shows that the dreaded disease is on the rise in the region (Singh and Kumar, 2010). In this report, the number of cases in Bhatinda, Faridkot, Muktsar district are limited only to the patients who came to the selected hospitals and, therefore, the sample itself is a very small portion of the actual victims. Being mostly illiterate, the victims often die without even knowing that an ailment like cancer exists. Even if they do come to know about it, the exorbitant costs and their inherent superstitions send them running to quacks. More than half the patients give up the treatment after one or two procedures. In such a dismal scenario, the ICMR estimates are probably just a fraction of the actual figures.

Table 4.11

*Distribution of Villages according to Extent of Flood
Caused by Ghaggar River*

Village	Total Land as per Records (Acres)	Area Damaged in Kharif Season in Acres				Maximum Damage (Acres)	Extent of Damage (%)
		1993	*1994*	*1995*	*2008*		
Ganota	548	110	-	98	222	222	40.51
Andana	4151	855	446	853	1680	1680	40.47
Hamirgarh	2770	540	130	650	1121	1121	40.47
Makorar Sahib	3850	504	664	1243	1558	1558	40.47
Baoupur	1744	586	480	626	706	706	40.48
Ghamur Ghat	1159	7	-	470	469	470	40.55
Banarsi	2755	1152	1259	1300	1115	1300	47.19
Mandvi	6088	3000	1952	2961	2464	3000	49.28
Fulad	1853	1030	525	800	750	1030	55.59
Rampura Gujran	1584	920	-	675	641	920	58.08
Bajidpur Nava Abad	801	485	338	365	324	485	60.55
Salemgarh	1957	1238	28	760	792	1238	63.26
Shahpur/Theri	870	500	516	578	352	578	66.44
Nava Gaon	3002	1250	1800	2120	1215	2120	70.62
Chandu	1275	-	895	903	516	903	70.82
Karhel	1742	1250	398	752	705	1250	71.76
Bhunder Bhaini	1043	850	82	400	422	850	81.50
Hotipura	1332	1050	800	1120	539	1050	78.83
Kundani	553	473	2	323	224	323	58.41
Bushera	2943	2605	980	2400	1191	2605	88.52
Khanauri Kalan	1678	540	374	540	679	679	40.46
Handa	1070	1020	476	1022	435	1022	95.51
Surjan Bhaini	617	573	513	590	250	590	95.62
Kabirpur	677	539	131	649	274	649	95.86
Banga	2738	951	1733	2700	1108	2700	98.61
Total	48800	22028	14522	24898	19752	29049	59.53
Moonak (Urban)	8063	5351	2490	5203	3263	5351	66.36
Grand Total	56863	27379	17012	30101	23015	34400	60.50

Source: Office of Deputy Commissioner, Sangrur.

Table 4.12

Number of Health Ailments in Relation to Chemical Consumption, 2004

State	Fertiliser Consumption		Pesticide Consumption		Per Capita NSDP		No. of Ailments/1000			
	Per Hectare	Rank	MT Technical Grade	Rank	NSDP	Rank	Rural	Rank	Urban	Rank
Andhra Pradesh	155.8	3	2133	9	20757	9	90	8	114	4
Assam	41.6	13	170	15	13139	12	82	9	83	9
Bihar	85.7	10	850	11	5780	15	53	15	63	13
Gujarat	106.8	8	2900	6	26979	4	69	11	78	10
Haryana	166.2	2	4520	3	29963	1	95	5	87	8
Karnataka	110.8	7	2200	8	21696	7	64	12	57	14
Kerala	67.4	11	360	14	24492	5	255	4	240	1
Madhya Pradesh	56.0	12	749	12	14011	11	61	13	65	12
Maharashtra	97.7	9	3030	5	29204	2	93	7	118	3
Orissa	40.4	14	692	13	12388	13	77	10	54	15
Punjab	192.5	1	6900	1	27851	3	136	2	107	6
Rajasthan	36.6	15	1628	10	15486	10	57	14	72	11
Tamil Nadu	152.9	4	2466	7	23358	6	95	6	96	7
Uttar Pradesh	125.5	6	6855	2	10118	14	100	4	108	5
West Bengal	129.0	5	4000	4	20896	8	114	3	157	2

Source: 1. *Statistical Abstract*, Punjab, 2004.

2. NSSO Report on Mortality and Morbidity, 2004.

Policy Dimension of the System

Recently, the state has come up with a policy initiative, 'Punjab State Water Policy (2008)' but that is at best just a beginning. Although water scarcity has been a serious problem for the last two decades, very little funds have gone into R&D for finding and putting into place the necessary techniques for the efficient use of water. This needs to be given top priority. There is a need to develop a long-term policy for groundwater use and recharge to maintain an optimum balance. The negative balance between annual available water supply and actual use needs to be corrected by using multipronged strategies like: (a) making maximum use of surface water; (b) increasing recharge; (c) addressing the urban sector needs and use and; (d) reducing/rationalising the demand for water. The state government passed the 'Preservation of Subsoil Water Ordinance' in 2008 to institutionalise delayed sowing of paddy. If Punjab is to continue as the food grains capital of India, modern agricultural practices will have to take into account the reality of the water situation and create a feasible plan for long-term sustainability.

Conclusion

The broad conclusion is that the Punjab model of agriculture has deteriorated the ecology of the region, in general, and the water resources in particular. The repercussions have started to show up in the form of depleted groundwater, widespread salinity, deteriorating water quality and specific kind of disease patterns in human beings. Clearly, over the years, a number of issues and challenges have emerged in the development and management of the water resources. The shortage of surface water availability, development and overexploitation of groundwater resources and deteriorating quality of water resources of the state have raised concerns on the need for judicious and scientific resource management and conservation. All these concerns need to be addressed on the basis of common policies and strategies with a vision of a new considered approach adopting emerging research in science and technology.

References

CSE (2005). *Analysis of Pesticide Residues in Blood Samples from Villages of Punjab, Report*. New Delhi: Centre for Science and Environment. Website: *http://cseindia.org/userfiles/Punjab_blood_ report.pdf*

Dung, N.H. and T.T.T. Dung (1999). "Economic and Health Consequences of Pesticide Use in Paddy Production in the Mekong Delta, Vietnam", *EEPSEA Research Report Series*. Available at: *http://www.eepsa.org*

Dung, Nguyen Huu, Tran Chi Thien, Nguyen Van Hong, Nguyen Thi Loc, Dang Van Minh, Trinh Dinh Thau, Huynh Thi Le Nguyen, Nguyen Tan Phong and Thai Thanh Son (1999). "Impact of Agro-Chemical Use on Productivity and Health in Vietnam", *EEPSEA Research Report Series*. Available at: *http://www.eepsa.org*

Gill, Karam Singh (2003). *Punjab Agricultural Policy Review*. Report for the World Bank. New Delhi.

Huan, N.H. and Le Van Thiet (2000). "Results of Survey for Confidence, Attitude and Practice in Safe and Effective Use of Pesticides", in *Agro-Chemicals Report* II(I), January-March 2002.

Jain, A.K. and Raj Kumar (2007). "Water Management Issues, Punjab-North West India". Website:*http://akicb.ifas.ufl.edu/upload/proceedings/jainak_water_ management.pdf*

Kishi, M., N. Hirschhorn, M. Djajadisastra, L.N. Satterlee, S. Strowman and R. Dilts (1995). "Relationship of Pesticide Spraying to Signs and Symptoms in Indonesian Farmers", *Scandinavian Journal of Work & Environmental Health* 21:124-133.

Leontief, W.W. (1963). *Studies in the Structure of American Economy*. New York: Oxford University Press.

Mathur, P.N. (1953). "An Efficient Path of Technological Transformation of an Economy", in T. Barna (ed.), *Structural Dependence and Economic Development*. Macmillan.

Pimentel, David, H. Acquay, M Biltonen, P. Rice, M. Silva, J. Nelson, V. Lipner, S. Giordano, A. Horowitz and M. D'Amore (1992). "Environmental and Economic Costs of Pesticide Use", *Bioscience* 42(10): 750-60

Singh, Karam (2007). *Punjab, the Dance of Water Table*. The Punjab Farmers Commission, Government of Punjab. p.13.

Singh, Karam and K.K. Jain (2002). *Dynamics of Structural Shifts in Cost and Returns in Farm Economy in Punjab*. Report for ACCP. Ludhiana: Agro Economic Research Centre, PAU. March.

Tardiff, R.G. (1992). *The e-pesticide Manual: A World Compendium*. 13th Edition. Alton, Hampshire.

Thakur, J.S., B.T. Rao, Arvind Rajwanshi, H.K. Parwana and Rajesh Kumar (2008). "Epidemiological Study of High Cancer among Rural Agricultural Community of Punjab in Northern India", *International Journal of Environmental Research and Public Health* 5(5): 399-407.

WHO (1990). *Public Health Impact of Pesticides Used in Agriculture*. New York, USA: World Health Organization.

5

Water Management Strategies in Punjab, India

A.K. JAIN

Introduction

Punjab (India), with a geographical area of 5.036 million ha (1.5 per cent of country's total geographical area), is a part of the Indo-Gangetic plains in the northwest of the Indian subcontinent and lies between $29^0 33'$-$32^0 31'$N and $73^0 53'$-$76^0 55'$E (Figure 5.1) Three perennial rivers, namely the Sutlej, Beas and Ravi, flow through the state. In addition, the Ghaggar, which is almost a seasonal river, flows through the southwestern part of Punjab. The water from the rivers is utilised for irrigation through a network of canal systems (14,500 kms) such as Sirhind canal, Sirhind feeder, Eastern canal, Upper Bari Doab canal, Bhakra canal and Bist Doab canal. The state has experienced a phenomenal increase in agricultural production during the last three decades, mainly due to extensive adoption of rice-wheat cropping system with assured irrigation facilities, leading the country in achieving food-sufficiency. Total food grains production in the state is 274 lakh tonnes. Punjab contributes about 55 per cent of the wheat and 42 per cent of the rice to the central pool.

The total population of the state is 277 lakh and about two-thirds of them live in rural areas comprising 12,278 villages. The annual rainfall (southwest monsoon) varies from 300-1,000 mm (Figure 5.2) with an average value of 530 mm and most of it (about 80 per cent) is received during July, August and September.

The seasonal rainfall does not come at fixed times, is not evenly distributed over the region and the amount is unpredictable. Thus, agriculture has to depend on other sources of water to meet crop requirements. With the expansion of irrigation facilities and rapid mechanisation of various agricultural operations, the cropping intensity has increased from 133 per cent in 1971 to 190 per cent in 2009 (Table 5.1).

Figure 5.1

Map of Punjab

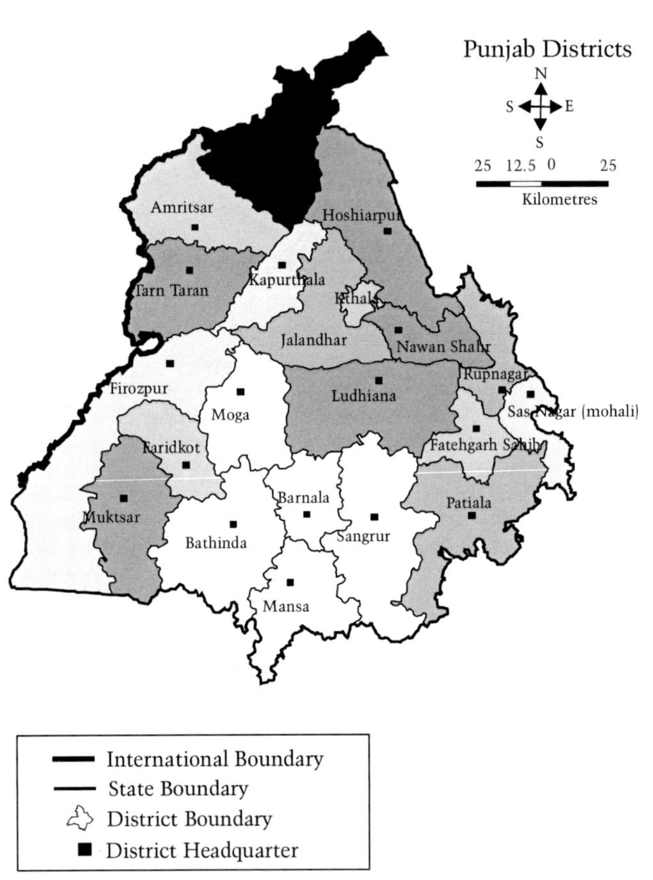

Figure 5.2

Average Annual Rainfall (2000-2008)

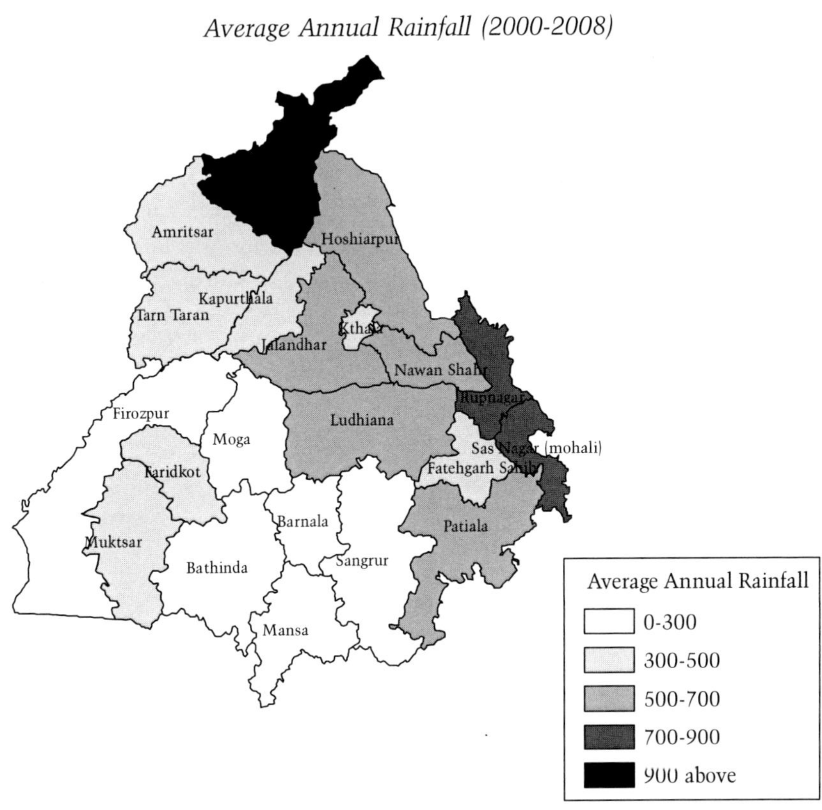

Table 5.1

Basic Agriculture Statistics of Punjab

Total geographical area	50.33 lakh ha
Inhabited villages	12278
Average rainfall	530 mm
Net area sown	42.01 lakh ha
Area under cultivation (%)	86
Area under irrigation (%)	98
Cropping intensity (%)	190
Area under rice	27 lakh ha
Area under wheat	35 lakh ha
Average yield	Rice-40 q/ha
	Wheat-45 q/ha
Tractors	4.2 lakh
Irrigation pumps	12.86 lakh

The state has 86 per cent crop area and 98 per cent of this is under irrigation that uses nearly 84 per cent of the state's water resources. Out of this, rice consumes 34 per cent, wheat 30 per cent and other crops 36 per cent. Out of the total irrigated land, 73 per cent is irrigated by tube wells and the rest by canal network systems. Most of the canal network exists in the southwestern districts of the state. The dominance of the rice-wheat cropping system has caused reduction of farming in areas where crops need less water. This has led to the overexploitation of groundwater resources, as the surface water is not adequate to meet irrigation needs.

The water demand of 3.65 million hectare metre in 1980 has increased to 4.76 million hectare metre. The total available water resources of 3.48 million hectare metre falls short of the total water demand (4.76 million hectare metre) by 1.28 million hectare metre. This deficit is met through overexploitation of groundwater. As a result, the number of tube wells has increased from 1.9 lakh in 1971 to 12.86 lakh in 2010.

Based on the groundwater statistics worked out by Directorate of Water Resources (Punjab), 110 blocks are overexploited, in which the groundwater withdrawal is more than 100 per cent, 3 blocks are at the critical stage where the withdrawal rate is between 90-100 per cent, 2 blocks are at semi-critical stage (70-90 per cent withdrawal) and 23 blocks are safe, with a withdrawal of less than 70 per cent. The safe blocks are located either in the foothill regions that has deep groundwater or in regions having poor quality groundwater. The groundwater quality status of Punjab is shown in Figure 5.3. The indiscriminate exploitation has led to the rapid depletion of groundwater resources in the state. The average groundwater exploitation is 166 per cent. During the last decade, the average fall of the water table in the state was 0.55 m/year. The general direction of groundwater flow is from northeast to southwest. The water table decline has a serious economic impact on the farmers, as they have to instal deep submersible pumps to draw water from ever deeper aquifers, resulting in greater energy consumption for pumping. Additionally, there is the water scarcity being forecast because of the climate change scenario, as also the competition from other sectors as a consequence of demographic growth, rapid industrialisation and urbanisation.

Figure 5.3

Status of Water Quality in Punjab

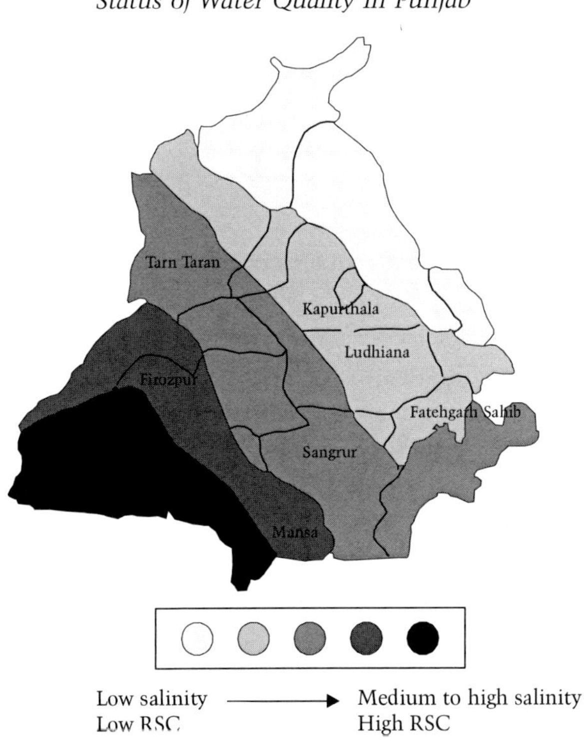

Water Management Issues

Based on the hydrologic and climatic conditions, the state can be divided into three zones—northeastern, central and southwest—that have varied water management issues.

Northeastern Zone

The northeastern part, covering an area of 0.95 million hectare (19 per cent of the state's area) includes the undulating foothills of the Shivalik ranges, known as Kandi area that occupies 10.6 per cent of the state's territory. The annual rainfall varies from 700 mm to 1,000 mm, but the precise amount and distribution is unpredictable. Rainfall-induced runoff and the ensuing soil erosion are major concerns. Almost 40 per cent of water is lost as runoff. Intense rains coupled with the undulating terrain and scant vegetative growth lead to heavy soil erosion that damages the agricultural land through the formation of rills and gullies. The quality of groundwater is good. Although aquifers are deep and soil rocky, these are now being exploited through community tube wells. Out of the total 37 blocks in this zone; 6, 17, 9, 4, 1 blocks have water table depth <5 m, 5-10 m, 10-15 m, 15-20 m and >20 m, respectively.

Central Zone

The central part, covering an area of 2.36 million hectare (47 per cent of state geographical area) is highly productive and is underlain with good quality groundwater. The rainfall varies from 400 to 700 mm. The water table is sinking because of overexploitation of groundwater resources to supplement the requirement of intense rice-wheat cropping sequence. The average annual fall in the water table was 0.09 m during 1974-1985. It increased to 0.20 m during 1986-2000 and further to 0.75 m during 2000-2010. Out of 74 blocks in this zone, 2, 10, 18, 22, 22 blocks had water table depths of < 5 m, 5-10 m, 10-15 m, 15-20 m and >20 m, respectively, revealing that in several parts of this zone levels have sunk very deep. This has forced the farmers to instal deep submersible pumpsets in place of centrifugal pumps, resulting in huge additional expenditure, extra power consumption and may even cause subsidence of land due to continuous overexploitation. With the large-scale installation of tube wells (two-third of the state's tube wells are located in this zone), almost 99 per cent of the cropped area has assured irrigation. Only 14 per cent area is irrigated with canal water supplies. The area with water table depth below 10 m increased from 3 per cent in 1973 to over 85 per cent in 2009. About 30 per cent area in this zone has water table depth of over 20 m. The cumulative fall in the level in this zone during the last three decades was over 9 m. It is also feared that there will be a reversal of flow from adjacent shallow water table areas (with poor quality water) towards this central zone if the declining trend is not checked immediately.

Southwest Zone

The southwestern part, covering an area of 1.71 million hectare (34 per cent of state geographical area), is recognised as the cotton belt of the state, but has groundwater quality problem that prevents its withdrawal.

The annual rainfall varies from 100 mm to 400 mm. Seventy per cent of the area is canal irrigated. The groundwater is brackish and has a high content of electrical conductivity (2.0 to 12.5 dsm^{-1}) and residual sodium carbonate (RSC) ranging from 3.0 to 35 mel^{-1}. High concentration of fluorine (F) has also been observed at a few locations. The area is affected with waterlogging, salinity and alkalinity problems, and has experienced rise in water level during the last two decades—*vis-à-vis* salt accumulation within the soil profile—mainly due to excess canal irrigation, inadequate drainage systems and lesser extraction of groundwater due to its poor quality. Out of 27 blocks in this zone, 8, 12, 5, 2 blocks have water table depth less than 5 m, 5-10 m, 10-15 m, 15-20 m, respectively. According to an estimate, about 75,000 ha of land covering 330 villages is affected by waterlogging. It has adversely affected the agriculture production and economic status of farmers. At some locations, a thin layer of good quality water lies over the brackish water.

Water Management Strategies

It is evident from the above mentioned zone-specific issues that effective water management in each region of the state is immediately required.

Management Strategies for Northeastern Zone

In the northeastern part, following strategies on a watershed scale are being adopted:

i) Controlling runoff water through the construction of storage tanks (water harvesting structures) and earthen dams offer a useful option in the Shivalik foothills area. The additional water helps in bringing more area under cultivation and also to provide life-saving irrigation to crops.

ii) Levelling of fields has been an effective erosion control and water conservation measure. Bench terracing is recommended on lands with slopes between 5 to 30 per cent. Wheat straw mulching greatly reduces the runoff losses and helps to conserve water in the fields.

iii) Water saving techniques such as optimal irrigation scheduling, irrigation methods (drip/micro sprinkler) should be adopted. In addition, crop varieties suited for rainfed cultivation need to be adopted to significantly reduce water requirements.

Management Strategies for Central Zone

The indiscriminate exploitation of groundwater coupled with reduced recharge due to urbanisation and growth of industries has led to fast depletion of groundwater availability. The following measures would be effective for regulating the groundwater withdrawal to sustainable levels.

i) The extensive rice-wheat cultivation has been the main cause of the rapid groundwater depletion. Therefore, efforts should be made to change, in some areas, high water requiring crop (rice) during the *kharif* to low water requiring crops (*basmati*, maize and pulses); and in *rabi*, some areas under wheat needs to be replaced with *raya* and chickpea.

ii) The adoption of on-farm water saving technologies would improve crop water productivity and increase efficiency in the use of water, thereby minimising groundwater use. The various on-farm water saving technologies that have been successfully adopted are listed below:

a) Laser levelling of fields (precision land levelling).

b) Timely transplanting of rice (low evaporative loss).

c) Optimal irrigation scheduling.

d) Growing of short/medium-duration varieties.

e) Efficient irrigation methods (drip/sprinkler/furrow).

f) Efficient water conveyance system (lining of water courses, underground pipeline system).

g) Mulching (moisture conservation).

h) Zero tillage (happy seeder).

i) Puddling rice fields (reducing percolation losses).

j)   Aerobic rice cultivation (direct seeded rice).

k)   Optimal plot size (as per soil type and stream size).

iii) The groundwater potential can be increased by recharging aquifers either with surface runoff or harvested rooftop water. The surface runoff available on grounds/roads/farms can be used for recharge of shallow aquifers. However, it should be ensured that the quality of this water is suitable for recharge. Rainwater harvested from rooftops being in general free from contamination, should be utilised by constructing suitable recharge structures.

iv) Sewage water after treatment can be used to supplement irrigation. Estimates are that sewage water can irrigate about 0.096 million hectare land annually with substantial contribution of nutrients.

v)   With the passage of time, there have been major changes both in cropping pattern and water requirement in different areas. So the canal water allowance/capacity factors/operational schedule needs to be revised to suit the present cropping pattern.

vi) The renovation of village ponds would increase the availability of water, which can be used for irrigation, as also to increase the percolation of stored water.

vii) There is a need to look into the possibility of virtual water trade through the import of water-intensive commodities to water scarce areas and vice-versa, for better management of available water resources.

Management Strategies for Southwestern Zone

The following management strategies could be adopted in the south-west zone having brackish groundwater.

i)   Cyclic/conjunctive use of canal water and tube well water is recommended for this area. Two irrigations of canal water followed by one irrigation with tube well water, or alternately mixing both the waters would reduce salinity to the level permissible for irrigation.

ii) If properly designed and constructed, surface drains will allow the speedier flow of rainwater, thus preventing waterlogging. These drains are low-cost structures and easy to maintain.

iii) Sub-surface drainage system that consists of a network of perforated PVC pipes buried below ground surface with suitable slope and an outlet could be laid to keep the water table at a pre-determined level.

iv) To tap good quality water floating over the brackish water, the multiple well-point system (skimming well technology) would prove useful.

v) Bio-drainage can be adopted in severely waterlogged and salt affected soils by planting eucalyptus.

vi) Fish farming in saline water would be economically rewarding in certain southwestern areas.

vii) Desalinised brackish water can be reused for secondary purposes.

Conclusion

The water management strategies for judicious use of available water resources in different zones of Punjab state have been discussed in the paper. In conclusion, groundwater irrigation will continue to be the major source of water, so there is need to implement these strategies effectively on regional scale. The researchers, extension workers and farmers should work together as a team that would help for sustenance of agriculture on a long term basis.

Section II

Running Dry: Micro to Macro Case Studies

6

Water as a Public Good *versus* Water Privatisation

UWE HOERING

The re-municipalisation of the water supply system of Paris at the beginning of 2000, until then one of the crown jewels of the French global players in the water sector, could be seen as a signal that the nearly two-decade-old heated debate on 'private *versus* public' has turned full circle. But the process of reversing privatisation—or to be more precise of private sector participation (PSP)—started already nearly a decade ago, when global water corporations like Suez/Ondeo, Veolia/Vivendi and Thames Water/ RWE announced their intention to reduce their engagement in southern countries.

This forced institutions like the World Bank to re-evaluate its privatisation strategy for the water sector, conceding that 'under current conditions the private sector will play only a marginal role in financing water infrastructure'.[1] Consequently, non-governmental, civil organisations and public utilities saw an opportunity to develop alternatives to privatisation. But first a brief look back on how it began.

The investment requirements in the water sector have been the central argument with which private sector participation has been promoted since the early Nineties. The expectation being that transnational private utilities would supply capital and modern management. More market, more competition and the entrepreneurial striving for profit would help remove the chronic problems many public utilities are faced with, such as high water loss and insufficient supply. This was the only way—so the mantra went—to achieve the Millennium Development Goal, i.e., to cut by half the number of people who do not have access to safe drinking water and appropriate sanitary installations by 2015.

---

1. "Water Resources Sector Strategy: Strategic Directions for World Bank Engagement", Draft for Discussion, 25 March 2002. p.38.

As a preliminary step, profound institutional and political adjustment processes were initiated to create positive investment conditions for private utilities in developing countries. The widespread habit of subsidising was replaced by the concept of 'cost recovery'. Private investors were encouraged with the help of various risk coverage instruments and by offering low interest loans for public-private partnerships (PPPs).

Since then, experience has shown that these projects contributed much less than expected to an improvement of the drinking water supply for the low income population, and even less so to an increase in the number of sanitary installations. This was even confirmed by a World Bank Report.[2] According to the report, even as privatisation results in an improvement in some cases, the basic problems remain: marginalised areas are hardly covered, corruption merely acquires a new shape, and accountability towards the public continues to remain weak. Frequently, privatisation has a negative effect on the poor, as in many cases prices have increased dramatically.

Nevertheless, despite these increases in water charges, corporations have had to concede that the expected easy profits in the water sector are not to be made, the main reason being that costs and returns in most areas of the water sector tend to be diametrically opposed. No wonder that J.F. Talbot, CEO of SAUR International, emphasised that 'the notion of cost coverage, particularly with regard to low-income groups, is untenable.'[3]

Private investments in many projects remained much smaller than hoped for or even agreed on during negotiations. One of the cases is Manila, where Suez/Ondeo has invested only a quarter of the capital that was originally promised. Instead of being supplemented by additional private resources, the investments continue to be financed by public means: by low interest multi- and bilateral development loans to governments that are then passed on to private implementing agencies.

Thus, the politics of privatisation creates a dichotomy in the water sector: lucrative areas such as the supply of drinking water for high income groups are transferred to private enterprises; less attractive areas such as

---

2. Harris, Clive 2003). "Private Participation in Infrastructure in Developing Countries", *World Bank Working Paper* No. 5. Washington D.C. April.

3. Speech at the World Bank in January 2002, *www.worldbank.org/wbi/B-SPAN/docs/SAUR.pdf*

squatter settlements, suburbs and rural regions remain with the public sector. This dichotomy corresponds with the dichotomy of public funds for the development of the water sector: on the one hand there is the promotion of the private sector and the minimisation of risk for global corporations, and on the other hand there are the alternatives that cannot be privatised, and where increasingly the poor themselves must become self-reliant to balance the lack of funds provided by the public sector.

The multifaceted political, economic and financial problems, however, with which the involved companies were confronted, turned out to be the basic problem facing the privatisation strategy.

- In many countries (Bolivia, South Africa, Indonesia and the Philippines) there was strong resistance against the water corporations, which, as in Cochabamba (Bolivia), led to a cancellation of the contracts.

- The financial crisis in Asia and the economic crisis in Argentina resulted in grave financial losses, especially for the second ranked of the global players, Suez/Ondeo. The devaluation of the Philippine peso and serious management errors resulted in the cancellation of the contract for Manila (West) which was at one time one of the World Bank's most prestigious projects.

- All three market leaders (Suez, Vivendi and RWE) accumulated large debts as a result of rapid expansion, which became a burden on the shareholder value; Veolia/Vivendi was up for sale after the collapse of the group.

Furthermore, corporate representatives conceded that 'low hanging fruit'—low risk projects that require little investment—have almost all been 'picked'.

Some corporations thus initiated a 'consolidation phase'. A central component of this consolidation was a retreat to—supposedly—secure markets such as the US, European countries with a low degree of privatisation like Germany, the Eastern European accession countries, or China. Still they claim that they cannot raise the investments necessary to achieve the Millennium Goals without considerable state subsidies and low interest loans. Thus, they are demanding a stronger engagement by the development banks—again with public money.

The World Bank and other multilateral and bilateral financial institutions and donors also became more reserved in their prognoses concerning the participation of the private sector in the countries of the South: 'We were too optimistic concerning the willingness to invest in these countries,' Nemat Safik,[4] the World Bank Vice President for Infrastructure, conceded, 'despite far-reaching reforms, many countries do not find investors.'

The experiences with privatisation and the decreased interest of water corporations also left its mark on a number of governments: 'Privatisation has not resolved the water problems for most of the population', is how Olivio Dutra,[5] responsible for urban planning in 'Lula' da Silva's first Brazilian government, summed it up.

The most obvious conclusion would have been to reorient towards an improvement of public utilities, which had been systematically placed at a disadvantage as opposed to PSP options. However, whenever reforms of public utilities were promoted within the scope of development cooperation, they usually served just as a preparation for privatisation, not as a means to improve the functioning of public utilities to remain public. The withdrawal of private global players could have been an opportunity for development cooperation to once again concentrate on public corporations as the central pillar of water supply and sanitation.

The picture has become much more diverse, mixed and differentiated than let's say a decade ago. Even as the drive for privatisation continues, the global players have set their sights on a more appealing target: countries with dwindling water supplies and ageing infrastructure, but better economies than developing countries. 'These are the countries that can afford to pay', says James Olson, a US attorney who specialises in water rights. 'They've got huge infrastructure needs, shrinking water reserves, and money.' Take China. Since 2000, when the country opened up its municipal services to foreign investments, the number of private water

---

4. Nemat Safik, cited in: *World Bank Pledges More Aid to Utility Projects Amidst Slump*. Washington. March 7 2003 (Bloomberg). *Link: http://lists.iatp.org/ listarchive/archive.cfm?id=68976*

5. Olivio Dutra, cited in: "European Water TNCs: Towards Global Domination?" in *Corporate Europe Observatory, Water Infobrief # 1* (no year of publication). Link: *http://archive.corporateeurope.org/ water/infobrief1.htm*

utilities has skyrocketed. But as private companies absorb water systems throughout the country, the cost of water has risen precipitously.

At the same time, there are many smaller, regional companies from emerging economies that are driving privatisation moves. This is the case in many countries across Latin America and Asia, less so in Sub-Saharan Africa.

In principle, the World Bank has not relinquished its privatisation strategy as can be seen from several strategy papers like the Water Resources Sector Strategy (WRSS) adopted in February 2003, or the Private Sector Development Strategy (PSDS) from early 2002, which focusses on infrastructure and services.

All the papers have two main ideas in common: (i) a widening participation of the private sector in the complete water sector and (ii) a rediscovery of large infrastructure projects. Similar strategy papers and political ideas also emanate from the Asian Development Bank, *viz.,* their agricultural sector programmes. By broadening existing instruments (guarantees, loans, etc.) and by developing new support measures (like output-based aid), the World Bank and other development banks are continuing to lower hurdles for participation by corporations in developing countries and making the investment conditions more attractive.

Nevertheless, the World Bank and other donors have become less enthusiastic about private sector participation even though they continue to promote it. Simultaneously, they now promote public water utility reforms, consumer corporations, and other non-private forms of management and ownership.

Additionally, the focus has shifted away from urban water supply and sanitation towards high dams and irrigation, where an increasing proportion of investments of the World Bank now go. Central to the new policy of the World Bank in the water sector is the development of a legal framework for water entitlements, the issuance of such entitlements, and the use of market-based mechanisms that permit voluntary adjustment by owners and users to meet temporary or permanent changes in demand. Investments in new or existing hydraulic infrastructure and irrigation projects are also considered, to provide a greater chance to introduce the basic concepts needed for the issuance of such water entitlements.

Thus, the focus shifts from the privatisation of infrastructure or management towards privatisation of the water resources itself. And water pricing has become the new magic formula, which has been the base of private sector participation since the 1990s: a higher water price is considered to bring about efficiency, investments and conservancy, which will also benefit the poor without access to water and sanitation.

Paris water is not the only example for re-municipalisation. There are many other prominent cases like Stuttgart and Berlin in Germany, Hamilton in Canada, Buenos Aires in Argentina, Dar es Salaam in Tanzania, or the move by the federal government of Malaysia, which is in the process of buying all water and waste water infrastructure in the country to develop them with public money. Instead of the so-called public-private partnership (PPP), there is a new model emerging of public-public partnerships (PuPs), where successful and experienced public utilities units team up with others to exchange information and experiences on how to improve public service delivery.

Most people involved in this process of reviving public utilities agree that merely a return to the conventional public provision utilities is no solution. Instead, there are several preconditions for success, drawn for example from cases like Porto Alegre (Brazil) and its concept of participatory budgeting. One of these is the participation of workers, employees and unions in the process, extended to participation of users and the public. Another is shifting of resources towards the public sector and the provision of public goods in spite of the precarious financial situation of many municipalities. Both preconditions point to the need and challenge for some fundamental shifts in policy and financial resource management, which are not easy to achieve.

This is not to argue that private sector and industry does not have a role to play. Or that there is no scope to make profit from investments in the water sector. With the right incentives, they can develop and supply the technology needed to make water delivery more cost-effective and environmentally sound. Ultimately, both public and private entities will have to work together. The question is: who shall be in the driver's seat? The answer depends on whether water is considered to be a common good and water supply a public responsibility, or not.

YU LI

7 Urban Water Challenge Today

A Case Study of Beijing

Since the early 1980s, Beijing has faced acute water shortages which have worsened even further following the impetus given to speed up the process of urbanisation. The resultant expansion and the increase in population concentration of the city magnified the problem of water security and water safety which, in turn, became a powerful restrictor of the urbanisation process, posing the biggest challenge to the efforts for the sustainable development of the cosmopolitan city.

Contradiction: High Population Density and Poor Water Accessibility

Beijing, the capital of the People's Republic of China, is the political and cultural centre of the country and also its international communications hub. It comprises 12 districts and two counties covering an area of 16,807.8 sq km.

According to a report in the *Beijing Evening News,* on July 22, 2010, the number of permanent residents in Beijing had reached 19.72 million, which far exceeds the figure in the population trajectory projected by the State Council for 2020. Going by the local statistics given in the report, the population of registered permanent residents in Beijing has reached 12.46 million and the registered floating population (i.e., those who have been living in Beijing for over six months) was 7.26 million. There appears to be an anomaly in the estimations of the two sources, with the state's projected population figures being significantly lower than that of the July 2010 report. The main reason behind this situation is the uncontrolled growth of the floating population, which, statistics show, had increased to about 15.18 million, with an average annual growth of 3.79 million during the past four years.

As in most regions of China, the water supply situation in Beijing remains grim. Its per capita water supply is as low as 285 cubic metres, which is 1/7th of the national average, and an alarming 1/30th of the global average. The per capita water access of Beijing ranks among the last 30 on a list of 130 key or capital cities of the world—far below the globally recognised bottomline of per capita water accessibility, which is 1,000 cubic metres. Beijing, thus, is a city of very high water stress,[1] much worse than the situation in Delhi.

Going by the 2003 data, the city's industrial, domestic, agricultural and environmental consumption rates were 23 per cent, 35 per cent, 39 per cent and 3 per cent respectively of the city's total water consumption.

The capital city's water resources have been overexploited for years, the ratio having risen from 67 per cent in 2000 to 76 per cent in 2003, directly causing an annual fall in the water level by 1.29 metres, seriously harming the water environment and the ecological balance. Four hundred million cubic metres of used water is drained into the river without proper treatment. Only about 40 per cent of the water is recycled back for consumption supply, while a huge quantity of clean water goes waste due to inefficient use. All this exacerbates the problems of environmental pollution and water scarcity.

With the rising living standards of the urban areas, the per capita water consumption has increased from 194 litres per person per day in 1980 to 316 litres in 2003, which means an annual increase rate of 2.2 per cent. In 2010, Beijing stated that the quantum of its water deficiency was nearly 2 billion cubic metres.

---

1. World Water Council: Water stress results from an imbalance between water use and water resources. The water stress indicator in this map measures the proportion of water withdrawal with respect to total renewable resources. It is a criticality ratio, which implies that water stress depends on the variability of resources. Water stress causes deterioration of fresh water resources in terms of quantity (aquifer overexploitation, dry rivers, etc.) and quality (eutrophication, organic matter pollution, saline intrusion, etc.). The value of this criticality ratio that indicates high water stress is based on expert judgment and experience. It ranges between 20 per cent for basins with highly variable runoff and 60 per cent for temperate zone basins. In this map, we take an overall value of 40 per cent to indicate high water stress. We see that the situation is heterogeneous over the world.

Water Sources

The gap between safe drinking water supply and demand in the capital is widening alarmingly, especially since the sources, the surface and underground waters, rely totally on rainfall. Given Beijing's location at 39°56' N and 116°20' E on the northwestern edge of the North China Plain and with the Bohai Sea about 150 km to its far southeast, the city hardly has any chance of enjoying an abundance of rain. It remains a semi-arid city with an average annual rainfall of 640 millimetre. In such a climate, the lakes of Beijing are of little consequence. Most of the water supply of Beijing comes from reservoirs, rivers and underground water, which amounts to a supply reserve of 4.76 billion cubic metres. Surface water, underground water and other water sources supply 24 per cent, 73 per cent, and 3 per cent, respectively, of the city's consumption. However, taking incidental or natural losses into account, the supply comes only to 1.972 billion cubic metres for the city's industrial and civilian consumption.

Severe Water Challenges

The eastern section of Beijing, which has been the focus of urbanisation is, for that very reason, the worst hit by the imbalance in the supply capacity and consumption demand, leading to severe water shortage. Both industrial and domestic consumption have been rising rapidly for the past 20 years, as a result of the special attention the government has paid to the development plans in the area. Not only has there been a sharp fall in the underground water level and increase in the salinity of the soil, but even the groundwater level has been sinking in this 1,000 sq km eastern area of Beijing.

Beijing has also been facing serious water safety problems for the past couple of decades—another fallout of modernisation. The situation is further exacerbated by the fact that most of the big wells are old and beyond repair; there is overdrawing of underground water as well as the improper drainage of used (sewage) water and garbage. Water pollution has become another baffling challenge to the urbanisation process. Of the 100 rivers of Beijing, there are 80 rivers under supervision. Statistics given by the authorities concerned show that 51 rivers, totalling 1,100 kilometres in length have been affected by pollution, of which 11 sections have suffered heavy pollution and 21 sections stand very heavily polluted. According to an estimation of the experts, the industrial and agricultural loss from water pollution reduces the annual GDP of Beijing by one per cent and, if we take indirect loss into account, by 3 per cent. Moreover, most of the pollution occurs when the rivers pass through the areas where the urban and rural conjoin, thereby polluting the water body as well. Unfortunately, this form of pollution continues even today, although the rate of contamination has somewhat slowcd down.

Means and Ways for Tackling the Problems

We have witnessed the authorities' ability to supply water touching bottom levels. What is of concern now is that the problems of water shortage and pollution pose a serious challenge not only to the development, but to the very survival of Beijing city if concrete measures are not taken to tackle the issue.

The first step would, quite obviously, be to increase the water supply to meet the demand. The Central government has already launched a variety of projects countrywide towards this end, such as constructing canals that would carry the waters from southern China to the thirsty northern regions.

Beijing's local government has carried out comprehensive measures, such as integrating the use of surface and underground waters, making more efficient use of rain and flood waters, feeding water back into the underground and regulating the supply. The positive results are already visible; the falling trend of the underground water level has improved and the initial foundation has been laid for the sustainable exploration of water, as also for promoting environmental purity.

The local government gives primary importance to measures for the economical use of water. With the aim of turning Beijing into an example of a water-saving city, the government has taken three major steps. Firstly, it focussed on rearranging the industrial structure, by setting up new and high-tech industries. Between 1980 and 2003, we have seen a decrease in industrial water use by 11 per cent on 10,000 RMB value. Secondly, it has set up watering mechanisms for the farmlands and adjusted the agricultural structures, thereby reducing the water consumption rate by 57 per cent between 1980 and 2003. Thirdly, laws and regulations are being revised for a better management of daily water use in the urban areas. There has been a decrease in the total amount of water consumption by as much as 27 per cent between 1980 and 2003.

Sewage water treatment and recycling of the treated water is an essential part of the water saving process. Beijing has a very low sewage treatment ratio—just 22 per cent. This is far from enough, considering the huge amount of liquid waste that could be put back to safe use. The local government has taken up a plan to build 30 more sewage treatment plants and conveyance systems to enhance its disposal capabilities.

Market areas will also find their place in this systematic project. The local government has already been considering reforming the water price regulation system. A survey is being carried out among Beijing residents to see if a system entailing different levels of prices for water would be acceptable to the payers. The survey so far has shown encouraging results and one can expect the new pricing system to see the light of day in the near future.

The local government must be very happy with the increasingly 'green conscious' social climate in Beijing. According to a recent survey, almost all Beijing citizens are aware of the serious water scarcity problem and 87.8 per cent of them recognise the urgent need to solve the problem. The survey also reported that nearly 70 per cent of the residents have pointed the finger at the huge amount of water being wasted, indicating that there is a need to intensify the efforts to save water.

A negative result of the survey is that only 4 per cent of Beijing families have access to recycled water, though 100 per cent of them favour the use of recycled water, either for environmental protection, or because it is cheaper.

More Areas of Concern

As compared to developed countries, China has a very low water environmental standard. There is no traditional, native reference dealing with the quality of water and the references today are those of the US, Japan, Russia, EU and Korea. China also needs to set up her own environmental evaluation system.

Adding to all the other challenges is the financial problem, restraining the hand of the local government from going ahead with environment protection and promotion programmes. All in all, China still has a long way to go to overcome this huge challenge.

8 Designing and Implementing Policies for Integrated Water Resources Management in Guyana

Pitfalls and Prospects

PAULETTE BYNOE

Introduction

Concern about global water security has led to a growing recognition and acceptance of the urgent need for all stakeholders or beneficiaries to prudently manage their water resources, given the current and predicted global climate change and its effects on this vital commodity. The Fourth Report of the Intergovernmental Panel on Climate Change (2007) confidently projects that, by mid-century, annual river runoff and water availability would increase at high latitudes (and in some tropical wet areas) and decrease in some dry regions in the mid-latitudes and the tropics. Additionally, many semi-arid areas (for example, in the Mediterranean Basin, Western United States, Southern Africa and Northeastern Brazil) are expected to suffer a decrease in water resources. Undoubtedly, water deficit in the 'South' will lead to prolonged droughts, reduced water availability, increasing demand for water by growing populations (especially in urban areas[1]), growing competition among users and increasing threats to water safety, which in turn can jeopardise the sustainable human development

---

1. The urban population is projected to increase from 47 per cent in 2000 to 53 per cent in 2015. Sixty-one million people are added to cities each year through rural-to-urban migration. Then there is the natural increase within cities, and the transformation of villages into urban areas. By 2025, the total urban population is projected to double to more than five billion, and 90 per cent of this increase is expected to occur in developing countries.

gains[2] that have been achieved over the last two decades. The result will be increased poverty and food insecurity.

On the other hand, many developing countries suffer from the delusion that water is 'plentiful and a free public commodity' for all to utilise without any established rules to guide its access, use and management. Yet, Chapter 18 of Agenda 21 of the United Nations Convention on Environment and Development (UNCED, 1992) underscores the need for protection of the quality and supply of freshwater resources by application of integrated approaches to the development, management and use of water resources. Additionally, the World Summit on Sustainable Development (Johannesburg, 2002) and the resulting Plan of Implementation (JPoA, 2002), in Article 26, called for all countries to: 'develop integrated water resources management and water efficiency plans by 2005, with support to developing countries.'

Water should be viewed from different perspectives: social, economic and environmental. It is therefore the responsibility of states and non-state actors to establish those 'rules' related to allocation, responsible use, conservation and environmental safety. This can only be realised by means of adequate financing and a realistic policy,[3] as also legislative and institutional framework that is still weak in many developing countries. In fact, an IDB Report (1998) notes that 'it is obvious to most governments that there is a need for policies and strategies for: controlling pollution; for establishing information bases on pollutants and water quality; to apply useful technology for pollution control and for waste treatment; to advance in institutional development and; to establish appropriate financing mechanisms.'

Theoretically, a National Water Resources Policy is intended for setting goals and objectives for the management of water resources and includes

----

2. According to the Organisation for Economic Cooperation and Development Policy Brief of February 2006 'Progress has also been made in developing countries, where between 1990 and 2000 access to safe water supply rose from 73% to almost 80% of the population.'

3. Although there has been significant institutional reform globally over the last decade, of 15 countries surveyed in the Caribbean, only three countries approved and adopted water sector policies; seven (including Guyana) are considering or pursuing such policies (Saleth, 2006).

policies for regional catchments, shared or transboundary water resources and inter-basin transfers—all best achieved within an Integrated Water Resource Management (IWRM) framework. It also deals with the management of the quantity and quality of both surface and groundwater resources and the delivery of water services. The translation of these policies into practical, on the ground, actions requires an enabling environment managed by legislation that is practical, enforceable and based on accurate scientific knowledge, economic incentives, the creation of institutions at the macro, meso and micro levels to ensure genuine local-level participation, monitoring protocols, financial and technical resources. These should be designed to ensure efficiency, equity and sustainability. Water policy will include matters of jurisdiction, delegation and issues such as the extent to which water management is decentralised or consolidated, the use of economic instruments, capacity building to meet institutional challenges, as also monitoring and control to reduce ecosystem degradation. Importantly, an estimation of its costs and benefits should be made, as the implementation of these policy measures would require substantial investments.

Water Resources in Guyana

Guyana lies 800 km south from latitude 88° N on the Atlantic coast to latitude 1° N and some 480 km east to west between longitudes 57° and 61° W. It is bordered by Venezuela on the west, Surinam on the east, Brazil on the south and southwest and the Atlantic Ocean on the north. It has a land area of 2,14,970 square kilometres and a population of approximately 7,67,000, which indicates an increase of 0.26 per cent from the pervious year.

In recent years, Guyana has been able to sustain a solid economic performance. The new GDP series, rebased on 2006 prices, shows that economic growth exceeded 4 per cent per year on average for the period 2007 to 2009. Nevertheless, approximately 30 per cent of the Guyana population lives below the poverty line. The minimum wage is currently about US $150 per month.

Basically, Guyana experiences an equatorial climate that is characterised by two wet seasons (May to mid-August and mid-November to mid-January) and two dry seasons (January to April and mid-August to mid-November). The average daily temperature is approximately 26.7 degrees Celsius. Relative humidity is high, with 80 per cent or more on the coastal zone, approximately 70 per cent in the savannah zone, and 100 per cent in the forested zone. Recently, the country has experienced weather anomalies associated with the El Nino/La Nina Southern Oscillation.

Four major landform types[4] can be identified in Guyana: (i) a low coastal plain that is relatively flat and accounts for approximately 6 per cent of the country's area and about 90 per cent of the human population; (ii) a hilly sand and clay region (south of the coastal belt) covering 25 per cent of the country and characterised mostly by vegetation cover, bauxite reserves and a sparse population; (iii) the interior savannahs which account for about 6 per cent of total land area is covered by grass, 63 per cent of the country's total land area and is marked by four major mountain ranges; and (iv) Precambrian land forms (for example, the Guiana Shield) and heavily weathered lateritic soils.

Guyana has a number of river systems, the main ones being the Essequibo River, the Berbice River and the Demerara River, each of which has several tributaries. Most of the freshwater requirement is met through seasonal rainfall, conservancies or aquifers. There is, however, an issue regarding the quality of this water. For instance, surface flows through the plains and sedimentary zones are often turbid, attracting high treatment costs.[5]

Groundwater, which consists of three distinct aquifers, provides about 90 per cent of the domestic water needs of the country. The groundwater system comprises three aquifers: the 'upper' sand, 'A' sand and the 'B' sand. To a larger extent, the country's demand for domestic water use is sourced from watersheds such as the Demerara Watershed[6] that serves the coastal population (accounting for approximately 90 per cent of the total population), particularly the Georgetown inhabitants. In the hinterland, water is obtained mainly from fresh water sources such as rivers and creeks.

---

4. Source: *http://www.guyanaguide.com/geog_reg.html*

5. See UNCED (1992).

6. The Demerara Watershed, drained by the Demerara River, encompasses sections of regions 3, 4 and 10.

There are increasing demands for water for various uses (irrigation, domestic, industry and commerce sectors), thereby challenging the availability of this resource. This competition is felt particularly in the dry seasons, during which severe water shortages are experienced throughout the country. The situation is aggravated by inappropriate water resource management and inadequate institutional arrangements. Uncontrolled water withdrawal, inadequate water tariffs and the absence of economic incentives for water conservation, all contribute to the wasteful use of water resources in both domestic and irrigation activities. Moreover, the environmental aspects of water development and urban sanitation are not given due consideration, resulting in water contamination (*National Development Strategy, 2001-2010*).

Added to this, the water resources, particularly watersheds, are compromised by environmental and human factors. The 1998 Report on Water Resources Assessment of Guyana notes:

> While abundant forest resources and forest utilisation have minimal direct impact on water resources, there are two areas that raise concern. One concern is the improper disposal of sawmill wastes, which raises biochemical oxygen demand levels and endangers aquatic life in the rivers. The other concern is the over-harvesting of forests in the White Sands area, which is degrading the timber stands to such an extent that they cannot regenerate. In turn, the reduction in forest cover could affect the recharge of the aquifer that provides most of the potable water for the country.

Bynoe (2008) confirmed the above in a *Situational Analysis Report on the Water Resources Management* in Guyana. The report itemises the principal threats[7] as: poor waste management practices by industries (sawmills, lumberyards and manufacturing firms); inadequate monitoring and regulation of sand, gold and bauxite mining; climate change and associated impacts such as temperature increase and flooding; unplanned expansion of human settlements in catchment areas leading to contamination of surface water, increased drilling of wells and; agriculture activities (farming and livestock rearing) and waste generation. Other threats cited are forest fires, unplanned tourism activities, poor waste management practices by households and industries (sawmills, lumberyards and manufacturing firms) and public misconception about and attitude towards water resources. The root causes lie mostly in institutional fragmentation

---

7. Based on the frequency of articulation of a particular threat by key institutions.

and inadequate policies, funding and institutional constraints that will be highlighted later in this paper (*National Development Strategy, 2001-2010*).

The Institutional Framework for Water Management

Currently, Guyana has no single national water policy. In fact, there are three major national policy documents that address, in part, aspects of water resources management. First, the Guyana Climate Change Action Plan (2001) outlines strategic measures to be taken to respond to the threat of climate change on water resources. These water conservation measures are expressed as: metering; the use of time-runs where the water supply may be staggered according to regions or sectors in the domestic/industrial sector; cautious development of new artesian wells in the interior regions for anticipated population influx from the coast; and the introduction of efficient control and management practices for water reservoirs network. Second is the Poverty Reduction Strategy Paper (2005),[8] which highlights several areas including 'improving the quality and delivery of services, ensuring good and effective regulation of the sector...mounting public awareness programs to educate families to conserve and treat water and rehabilitation and maintenance of water infrastructure.' Finally, the National Action Plan for Combating Land Degradation (2006) identifies specific actions to be taken in respect of watersheds in Guyana, which can be listed as: promotion of and support to sustainable management in forest and mining; restoration and protection of biodiversity and watersheds in collaboration with relevant agencies; undertaking of an groundwater situation analysis including an assessment of salt water intrusion; undertaking of a study and situation analysis of both coastal and hinterland surface water systems and; the expansion of the Water User's Association and their role in management.

In an effort to reform the water sector, the Government of Guyana enacted the Water and Sewerage Act 2002 'to provide for the ownership, management, control, protection and conservation of water resources, the provision of safe water....' The Act 'governs all water rights and facilitates the introduction of national water standards and a National Water Council to spearhead the water resource management policy.' Specific issues

---

8. For more information, please see *http://www.povertyreduction.gov.gy/index.html*

covered are water supply and connection, water regulations, wastewater and sewerage matters, drought orders and hydrometeorological matters. Importantly, this Act establishes a new regime for water use and management in Guyana. Section 18 of the Act declares that 'the ownership of all water resources and the rights to use, abstract, manage and control the flow of water are vested in the State.'[9]

Other pieces of legislation which predominantly address impact issues include the Environmental Protection Act, No. 100 of 1996, the Mining Act (2005) and the East Demerara Water Conservancy Act (1998). The Environmental Protection Act (1996) provides for 'the management, conservation, protection and improvement of the environment, the prevention or control of pollution...and the sustainable use of natural resources'. It is the umbrella legislation that mandates the undertaking of a number of measures to safeguard the environment and its resources, including water. For example, Part IV (19) 1 states, 'A person shall not discharge or cause or permit the entry into the environment of any contaminant in any amount, concentration or level in excess of that prescribed by the regulations or stipulated by any environmental authorisation.' It should be noted that although Water Quality Regulations were formulated in 2000, the point of reference for water quality standards remains the Pan American Health Organization (PAHO)/World Health Organization (WHO) standards.

As stated earlier, mining activities are a major threat to water resources in Guyana; as such, the Mining Act (2005) seeks to prevent actions that lead to the contamination of rivers, creeks and other waterways—the habitat of various life-forms. These actions encompass mercury use, mine reclamation, mine effluents, contingency planning, mine waste and tailings management. The law strictly prohibits: the use of mercury during primary stages of mining operations such as in sluice boxes, hammer mills or ball mills; discharge of amalgamation tailings (black sand or fluids which contain mercury) into water bodies; discharge of fluids in excess of 30 nephelometric turbidity units (NTU) or 100 total suspended solids (TSS); burning of amalgam in open air and making ponds less than 20 metres away from rivers or other waterways. Further, the

---

9. Janki (n.d.).

Mining (Amendment) Regulations Act 2005 cover, among other issues: the use and handling of mercury; the use of retort for burning of amalgam; the use and handling of cyanide; the provision of drinking water; registering of poisonous substances and storage of poisonous substances. The East Demerara Water Conservancy Act (1998) seeks to establish the East Demerara Conservancy Board for the purpose of making better provision for the supply of water to the inhabitants of Georgetown, and to provide for the management of the Conservancy through the establishment of a Board of Commissioners.

Currently, at least six entities are tasked with the management of water resources in accordance with their legal mandates: Guyana Water Incorporated (GWI), Guyana Sugar Corporation (GuySuCo), the Ministry of Agriculture Hydrometeorological Service Unit, the National Drainage and Irrigation Authority (NDIA); the Environmental Protection Agency; and the Guyana Geology & Mines Commission (GGMC). An example of institutional responsibility can be cited in the Guyana Water Inc. (GWI),[10] a merger of two water utilities established in 2006 to control, maintain and manage the sewerage system and waterworks of Georgetown. The GWI has been tasked with the management of the system and to deliver water to the suburban, rural and the hinterland regions, excluding Linden and those areas supplied by the Sugar Industry Labour Welfare Fund Committee. This entity was appointed by the Minister of Housing and Water as the public supplier who has the duty to provide potable water for domestic purposes and a satisfactory supply of water for industrial and commercial purposes. Initially, the GWI was managed by an international consultancy firm, but the Government of Guyana terminated the contract and reassumed management responsibility due to the continued deteriorating water and sewerage services.

The Hydrometeorological Service Unit, on the other hand, has the legal responsibility for managing the licensing system for groundwater and, when it comes into operation, the system for licensing surface water. In addition, the unit is responsible for verifying existing lawful uses and bringing those

---

10. GWI is a public company incorporated under the Companies Act, 1991 with the Government of Guyana as the sole shareholder; it is licensed by the Ministry of Housing and Water (MH&W). The company provides water to approximately 164,000 homes, offices and schools across Guyana.

uses within the regulatory framework established by the Water and Sewerage Act 2002, and for establishing and maintaining systems to monitor water quality and use. Interestingly, the unit is accountable to the Minister of Agriculture and not the Minister of Housing and Water (Janki, n.d.). Undeniably, the realisation of the goals of the integrated water resources management is contingent upon institutional collaboration, integration of working programmes and plans of action, and above all an institutional arrangement that will facilitate such collaboration and integration. However, what has been discussed above underscores a number of pitfalls in the institutional framework that will be discussed below.

Pitfalls in the Institutional Architecture of Water Resources Management in Guyana Legislative Framework

Currently, the various pieces of legislation related to water resource management in Guyana are not effective since they are compartmentalised due to the existing fragmentary organisational framework. Additionally, there is no set policy on watershed management and enforcement is very limited due to limited monitoring and enforcement capacity of key institutions, which are related partly to institutional constraints.

Fragmentation

In the absence of a functioning single entity—the National Water Council in 2008 under the Water and Sewerage Act 2002—various individual entities pursue their own organisational mandates, priorities, programmes and projects without any holistic perspective on water resources management issues. Such a fragmentary approach and lack of integrated planning leads to increasingly competitive demands and conflicts among users, as witnessed in Guyana during the 2009/10 El Nino dry spell; it resulted in a conflict of opinion as to whether water should be treated as an essential resource for 'economic good' (GuySuCo) or as a vital item for 'social good' (GWI). Efforts have been made to resolve such conflicts at other levels, for example, by the East Demerara Conservancy Board. Presumably, water resources have been regarded as an issue of production and consumption, rather than that of integrated management which would also include its users and links with other water uses (Corredor, 1996 in IDB, 1998).

However, an effective National Water Council should result in cross-sectoral linkages, greater integration of work programme activities, reduced duplication, sharing of information, technical assistance and, ultimately, a more effective management regime for resources management in Guyana. This is contingent upon the provision of adequate financial resources to support the work of the Council.

The formulation and implementation of a national water policy is contingent upon an effectively functioning National Water Council that has the onus of developing and providing advice on the incorporation of the national water policy into plans, programmes and activities of various entities, as well as managing and coordinating all activities related to the policy, among other things.

There is also need to consolidate and harmonise the various pieces of legislation that govern user issues, such as extraction, use and conservation of water resources, as well as impact issues in terms of reducing the negative effects of human activities. The current fragmented approach to the enactment of legislation undermines the effectiveness of the measures already enforced. A typical example is the Environmental Protection Act of 1996. The enactment of the key piece of legislation should have resulted in a review of other pieces of legislation to avoid possible conflicts, duplication or overlaps. Importantly, the Water and Sewerage Act needs to be supported urgently by regulations to become effective. Related institutions also need to strengthen their enforcement mechanisms to give effect to the existing legislation of water resources.

Unrealistic Water Tariffs

In Guyana, tariffs are fixed by the Public Utility Commission—an institution that has limited financial and human resources to perform its duties effectively. Moreover, since it was established in 2002, the GWI has not attempted to restructure its legacy of 27 separate tariff categories, barring adjustments of the rate increases. National legislation allows the GWI to introduce specific tariff structures which are not necessarily rational. There are four major structures, namely residential,[11]

---

11. This is for urban and rural communities that the Government of Guyana considers to be occupied by the poorest of the poor citizens and will subsidise payment of water bills.

commercial,[12] industrial[13] and institutional.[14] Residential utility is further sub-divided into five differentiated bands with annual per capita payments amounting to less than US $150.[15] Persons who receive treated water pay a slightly higher tariff. On the other hand, the criterion used for the latter three structures is the size of the building in square feet—a very crude way of determining use. At the household level, particularly in the urban areas, the water entity has introduced the metering system, which is superior as a deterrent to inappropriate behaviour exhibited by households. This must be understood from the perspective of cost recovery, given that the sector relies heavily on Government subsidies: the state provides approximately 40 per cent of the cost of operations, mostly related to the cost of energy. As IDB (1998) has pointed out, the delivery of water services is typically centralised in government organisations and agencies which are often too overextended, underfunded and ill-organised to provide quality service, resulting in deteriorated infrastructure and low efficiency. Cost recovery is extremely important, especially since Georgetown's population has increased by 500 per cent. But the flawed public perception, supported by a legislation that keeps the price of water at a minimum in the interest of public health, equity and (possibly) political patronage, has not helped to instil efficiency in the use of the resource. Consequently, most water utilities do not cover their operating costs and are often saddled with low levels of bill collection and high levels of corporate debt (FAO, 2010), which was, for a long time, the sole domain of the public sector.

---

12. This is applicable for any premises where commercial activity is carried out. Includes but not limited to premises where the activity carried out does not depend on the use of water. Example, dry goods store.

13. This is applicable for any premises where manufacturing activity is carried out. Includes but not limited to premises where the activity carried does not depend on the use of water as an ingredient.

14. This is applicable for any premises where non-domestic activity is carried out. This includes but is not limited to premises where the activity carried out depends on the use of water for sanitary purposes. Example: private school/sports club with less than 20 persons meeting no more than once weekly.

15. In general, approved tariffs in Guyana are somewhat lower than those observed in other Caribbean countries, but the ability of customers in Guyana to pay these approved tariffs is also lower, due to lower income levels. GWI has two main sources of income—tariffs and Government subsidies. In 2007, there was a gap between GWI's income and expenditures: in 2007 even with a significant GoG contribution of $1,117 million, GWI's expenses exceeded its income by $660 million after making an allowance of nearly $1,185 million for depreciation.

There still remains the above mentioned issue to be resolved. If water is treated essentially as a resource of 'social good',[16] it would have implications that go counter to the efficient use of water as a factor of production and conservation; treating it as a source of 'economic good' would undermine the task of poverty alleviation.

The agriculture and manufacturing sectors should understand that water cannot be an issue that is divorced from water resources management. Fundamentally, the purpose of making people pay for using water are: to recognise water as an economic good and give the user a sense of its real value; to encourage the rationalisation of water use and to raise revenue to finance the programmes and interventions provided for in the water resources plans.

Failure to charge[17] a price for water to cover the real cost is what creates those circumstances—too often repeated in the world—which underlie and are the cause of the current concern with scarcity (Lee, 1997). Given the problem of affordability, policymakers should pursue other opportunities for creating additional revenue for the water sector, including enforcement of legislation regarding pollution and environmental charges.

Monitoring and Enforcement

The pressure to promote economic and social development in Guyana has given rise to a number of large-scale projects, such as infrastructure development linked to broader national development projects, natural resources exploitation (both surface and sub-surface), hydropower development, as also mining and logging activities. The apprehension of pitfalls likely to be caused by these projects is rooted in the absence, to date, of scientific data on water resources—a critical tool for decision-making—coupled with the lack of institutional capacity to monitor and regulate activities affecting watersheds. Effective capacity in regulation and

---

16. The quality of water provided also does not meet the targeted WHO water quality standards. Iron concentrations generally exceed 0.5 mg/l and lead to discoloration of clothes and plumbing fixtures and fittings. Poor water quality has also affected meter operation where these have been installed.

17. One of the goals of the GWI is financial self-sufficiency: achieve financial self-sufficiency through increased tariffs so that income meets operation and maintenance costs by the end of year five (5) and operation, maintenance and depreciation costs by the end of year ten (10).

enforcement is critical. Moreover, the strong motivation that drives the titled hinterland communities to get involved in logging and mining activities (an entitlement under the Amerindian Act, 2006), as well as the ongoing legal and illegal logging activities in the Linden-Soesdyke area by persons who have made charcoal[18] production their dominant livelihood necessitates effective monitoring and enforcement regimes—both of which are currently inadequate at the Hydrometeorological Service Unit and the Environmental Protection Agency. Similarly, the depth of sand mining in the hilly sandy area (recognised as the 'rechargeable' area for coastal water supply) need to be carefully monitored to prevent any possible contamination of aquifers. Monitoring necessitates decentralisation which is virtually impossible without the availability of adequate financial and human resources.

Political Commitment

Integrated water resource management requires the commitment/support of all key decision-makers in Guyana. The current reality is that the ministries/agencies just do not have the required will or vision to promote the efficient and safe use of the water resources, as reflected in the fact that each sector takes its own independent decision on the issue. Commitment on the part of sector ministries/agencies must be expressed in budgetary allocations and funds to support activities aimed at promoting the efficient use of water resources by all segments of the population. In point of fact, most of the initiatives regarding water resources management in Guyana are donor-driven (the United Nations Development Programme, the Inter-American Development Bank and the European Commission) and this, in essence, prevents the cultivation of a more holistic view of water resources management.

Limited Institutional Capacity

A key challenge of institutions with direct or indirect responsibility for water resource management is the limited capacity, in terms of human resources (particularly qualified[19] and motivated staff), equipment for

---

18. Charcoal production is a traditional livelihood activity.
19. More than two-thirds of the University of Guyana graduates migrate to other countries in the Caribbean, North America and Europe.

monitoring and financial resources to execute various activities that promote integrated water resource management. For example, the Hydrometeorological Service Unit is the regulatory body for water resources management, but has very limited capacity to carry out its mandate, including data collection on groundwater resources. According to a 1998 Report on Water Resources Assessment of Guyana, 'Hydrologic data are lacking throughout the country, particularly since the late 1960s when data collection decreased dramatically.' In fact, with only two hydrologists, it is almost impossible to monitor water resources over a land area of 83,000 sq miles.[20] Today, the issue is still relevant. Water resources management stands handicapped by the lack of adequate and reliable hydrologic, meteorological, and water quality data, as well as information on socioeconomic characteristics and indicators of water use efficiency and, in general, reliable indicators to be used as a basis in conflict resolution. Additionally, there is the challenge of adequate and timely scientific data upon which informed decisions can be made. The paucity of data is affected by the human resource capacity, which is a function of the current remuneration package by the Ministry of Agriculture.[21] A related challenge is the need for institutions to exchange available data on water resources. National institutions seem unaware of previous studies that have been done on water resources or the 'location' of study reports.

Absence of a Water Conservation Ethic

The public, being poorly informed, continues practices which result in significant wastage of water. This absence of water conservation ethics cannot be overcome by mere public relation exercises. There is a dire need to inculcate awareness about the issue through a comprehensive public awareness and education programme which utilises both formal and non-formal approaches. The time has come for policymakers to recognise the powerful role environmental education can play in creating a responsible citizenry and in fostering greater cooperative efforts among water users, particularly farmers and miners.

---

20. This can be interpreted as a logistical problem.

21. The Hydrometeorological Division is unable to retain highly qualified staff due to unattractive salaries and benefits.

National Development

As noted above, in its efforts for socioeconomic development, Guyana has launched a number of large-scale projects encompassing the fields of infrastructure, natural resources exploitation (mining and logging) and hydropower development. These projects together with climate variability and longer-term climate change have given rise to new challenges as a result of the highly insufficient scientific data on water resources, the lack of institutional capacity to monitor and regulate activities affecting watersheds and the uncontrolled exploitation of land.

Policy Prospects

Guyana can overcome these challenges by exploiting, both now and in the future, the opportunities that present itself for integrated water resource management. These potential solutions are given below.

Establishment of a National Water Council, Development of a National Water Policy and Roadmap

The establishment of the National Water Council in Guyana was a very positive step towards promoting protection of water resources. The two key functions of the Council is to develop and/or review the national water policy and to oversee its management and coordination—the latter being essential for the avoidance of overlaps and duplication of tasks, as also for keeping a tight rein on competition and conflict amongst the water users. One of the goals of the policy is to provide a framework to maximise the contribution of the water sector to sustainable economic, social and environmental development in an efficient and equitable manner. The Council precedes the policy; hence the urgency of ensuring that this governing body becomes fully functional.

In accordance with the Water and Sewerage Act 2002, the National Water Policy provides for, *inter alia*, more efficient use of water resources by all users, as well as environmental protection and pollution control in cooperation with other institutions such as the Environmental Protection Agency and the Ministry of Agriculture. It is envisaged that this policy will set out strategies, objectives, plans, guidelines and procedures to ensure the following: (i) equitable allocation of water for the social and economic benefit of all Guyanese; (ii) management of water resources for aiming at sufficiency and sustainability; (iii) protection of communities from severe hydrological events (for example, El Niño); (iv) protection and conservation of surface water and sustainable use of groundwater and (v) recognition and protection of the existing rights. The implementation of this policy is of paramount importance for putting in place an effective instrument for the overall management of the nation's water resources.

Land Use Policy and Sustainable Land Management

When fully reviewed and implemented, the Land Use Policy will provide an overarching framework within which a National Land Use Plan can be created to ensure the optimum and rational use of land, as also the integration of its various uses (GoG, 2007).

Meanwhile, Guyana, under the Sustainable Land Management (SLM) Project[22] is benefitting from a number of projects that seek to collectively 'establish an enabling environment to combat land degradation through a participatory process of capacity building; mainstreaming of SLM into national development strategies and processes; broad stakeholder participation and resource allocation for SLM.'[23]

To date, two programmes have been implemented. First, a training seminar was held to provide decision-makers with an assessment of the nature and extent of land degradation in Guyana, and for enhancing the capacity of a cadre of Guyanese technicians to conduct degradation assessments. The second focussed on water analysis and management. Noteworthy are two other projects that will be executed shortly. The first is a Resource Valuation programme for building capacity in the valuation of natural resources, so as to enhance policy analysis and decision-making. The second project deals with Early Warning Systems; it aims to provide decision-makers with a comprehensive, analysis-based set of options for the introduction of early warning systems appropriate to the Guyana situation, while at the same time integrating the local knowledge input with the system.

---

22. The project is financially supported by the Global Environment Facility (GEF) and implemented jointly by the Guyana Land and Surveys Commission (GSLC) and the United Nations Development Programme (UNDP).
23. Information taken from UNDP website: *http://www.undp.org.gy/project 00047476.html*

There is also a regional project, 'World Wildlife Fund (WWF)—Guyana's Sustainable Natural Resources Management Project 2007-2011', which includes the study of freshwater conservation and management. Information provided by the WWF indicate that the project will promote conservation of freshwater ecosystems through necessary legislations, strengthening of the concerned institutions and the introduction of improved water management systems in gold mining areas to reduce mercury contamination.

International Conventions

Guyana's obligations to major international conventions such as the United Nations Convention on Biological Diversity (UNCBD), the United Nations Convention to Combat Desertification (UNCCD), and the United Nations Framework Convention on Climate Change (UNFCCC) provide excellent opportunities for the country to access the much needed financial resources and technical assistance from the world bodies so that it is enabled to respond to the various challenges it faces in the water resources sector. Given the global focus on climate change and the existence of ecosystem services, Guyana can confidently lobby for support in this area. Further, the nation's entry into the Cartagena Convention on land-based pollution and point source pollution will help complement the government's legal initiatives, provided Guyana has the necessary capacity to implement the regimen of the Convention.

Research, Information and Knowledge Sharing

Information and knowledge sharing regarding watershed conditions and management issues among all jurisdictions and agencies that are responsible for water resources is essential for effective management. Besides, such sharing is unavoidable, given the costs involved in acquiring fully reliable data from geographic information providers, such as through satellite imagery, etc. A blanket sharing of all information becomes a matter of necessity—not of choice.

The Guyana Water Inc. (GWI) has recently established a Water Resource Unit that focuses on use and monitoring of extraction of the resource. This initiative will facilitate data collection and will provide vital information to the concerned agencies.

Guyana is also benefitting from two projects aimed at capacity building.

i) Increasing the Capacity of Hydrological and Meteorological Monitoring Networks as a Means of Tracking Climate Change and Rise in the Sea Level

Guyana's National Hydrological Stations Network (NHSN) is currently performing far below its required capacity, both in terms of infrastructure and human resources. The result is that data collected by the Hydrometeorological Service is too meagre to allow for a comprehensive understanding of the country's hydrometeorological status.

The objectives of this two-year capacity building project are to enable the NHSN and the NMSN so as to achieve the following targets.

- To collect reliable and comprehensive hydrological and meteorological data.

- To track global climate change and sea level rise to serve as a global model for tracking to be applied in countries with geographic conditions similar to Guyana.

- To involve stakeholders in monitoring national climate, thereby allowing them to better appreciate climate variability in the long term and serve as a global model demonstrating how local people and communities can be involved in this field.

The project would be implemented by the Hydrometeorological Service, Ministry of Agriculture in collaboration with the concerned agencies and NGOs, including the Iwokrama Conservation International, the Guyana Marine Turtle Conservation Society, as also the communities involved. Funding for project implementation would be sought from GEF Amazonia's project and the Amazon Cooperation Treaty Organization.

ii) Collaborative Agreements between Hydrometeorological Service Unit and Other Stakeholders

This project seeks to formulate agreements for coordination amongst the Service, other agencies, NGOs and local communities for data collection. Its activities would include working out collaborative arrangements for hydrological monitoring by the Hydrometeorological Division, other agencies, NGOs and local communities, by sharing of the collected data among the partners and discussions on capacity building issues. The expected outputs are: formulation of collaborative data collection agreements; the building of capacity required for effective and efficient data collection; the involvement of local people and communities in collecting, analysing and interpreting data and setting an example of how local people can be positively involved in the above mentioned tasks— a documented model that would be a useful lesson for other Amazonian states, as also the global community.

The GWI can establish partnerships with the University of Guyana (which currently offers an MSc in Water Resources Management) to build capacity in the area of base-line and other studies aimed at enhancing the process of decision-making.

Establishment of the National System of Protected Areas

The National System of Protected Areas aims to delineate at least six protected areas, these being the Kaieteur Falls area, Kanuku Mountains, Mount Roraima, Orinduik Falls, Shell Beach and the South Eastern Forests.[24] This safety net will no doubt protect the country's forest cover and, by extension, help maintain a constant supply of good quality water. To date, Guyana has three legally protected areas—the Kaieteur National Park, the Iwokrama Rainforests and Shell Beach.

Strategic Environmental Assessment

Strategic Environmental Assessments (SEAs), which originated from land use planning in the developed world, provide a useful and an effective approach to managing the diverse uses of water resources and preventing or reducing conflicts. In essence, the SEA process will allow decision-makers to focus on key issues, which are: different spatial scales; multisectoral decision-making; stakeholder participation; monitoring, evaluation and broadening of the perspectives beyond immediate sectoral issues—all of which are vital for integrated water resources management.

---

24. *http://www.epaguyana.org/npas/proposed.htm*

Partnership and Collaborative Management

Governance is a more inclusive concept than government *per se*; it definitely embraces the relationship between a society and its government (Rogers, 2002). In keeping with the Dublin Water Principles (1992),[25] the Hague Ministerial Declaration on Water Security (1998) called for 'governing water wisely to ensure good governance, so that the involvement of the public and the interests of all stakeholders are included in the management of water resources.' At the Bonn 2001 Ministerial Declaration, the ministers recommended action in three areas and named water governance as the most important among them: they proposed that 'each country should have in place applicable arrangements for the governance of water affairs at all levels and, where appropriate, accelerate water sector reforms.'

Effective management of water resources in Guyana requires public participation, i.e., collaboration among state entities, the private sector and local communities. The National Water Policy can create opportunities for the creation of partnerships among key stakeholders of water management, as well as specify specific roles that will demonstrate a genuine participatory process.

Role of Donors and the International Community

Donor institutions can drive policy reforms and build technical capacity in the water sector in Guyana. For example, the IDB Integrated Water Resource Management Strategy for Latin America and the Caribbean promoted a strategic shift from development to management and from a sectoral to integrated problem solving. This followed the acceptance of principles formulated at the Dublin Declaration, Agenda 21, the San Jose Declaration and Action Plan and approved by the Heads of State at the Summit of the Americas on Sustainable Development held in Santa Cruz de la Sierra, Bolivia, in December 1996. In respect to capacity building,

---

25. The Dublin principles that guide the Integrated Water Resources Management (IWRM) principles are: (i) Fresh water is a finite and vulnerable resource, essential to sustain life development and the environment, (ii) water development and management should be based on a participatory approach, involving users, planners and policymakers at all levels, (iii) women play a central role in the provision, management and safeguarding of water and (iv) water has an economic value in all its competing uses and should be recognised as an economic good.

donor institutions can collaborate with the University of Guyana and the University of the West Indies to offer professional courses using a multimedia approach. This type of capacity building for integrated management must be sustained in order to meet the capacity needs of those agencies that have varied responsibilities, but with a common vision for water resources management in Guyana.

Recommendations

The following recommendations can be made in the context of the aforementioned issues.

- The National Water Council should invariably constitute a Technical Inter-Agency Group that is tasked with providing technical support for watershed management in Guyana, which include representatives of women, the private sector and farmers.

- Priority should be given to the completion of the National Land Use Plan and the National Water Policy and adequate resources should be made available to ensure prompt implementation.

- Since water cuts across all productive and social sectors of the economy, policymakers should consider taking urgent action to mainstream water resources management into national socio-economic development plans.

- Community members should be trained so that it empowers them to participate in watershed management and water resource protection, particularly because of the capacity constraints faced by key institutions.

- There is an urgent need to establish a Monitoring and Enforcement Unit within the Hydrometeorological Division and to provide adequate resources to facilitate its operations.

- A central clearing house for information on water resources should be established with adequate human and financial resources to achieve its mandate and strategic objectives.

- A comprehensive national education programme should be launched to inculcate an individual and collective sense of responsibility for water management among Guyanese societal groups.

- A Strategic Environmental Assessment should be conducted for the water sector in Guyana.

- The Government of Guyana should give further consideration to the decentralisation of water resources management, for example, by empowering grassroots women organisations and community-based development groups to play a more visible role.

- Once completed, the national water policy, together with a roadmap, should be accompanied by an integrated water resources management plan that would facilitate the implementation of the new water resources policy. There is a need for a roadmap to clearly identify resources, and a chart to clarify the roles of institutions, as also an element of monitoring and evaluation component.

- A sustainable financing system must be developed with inputs from the public, private and international stakeholders. There is also a need to review the current differentiated structures for water tariffs. As posited by San Martin (2002), 'increased awareness on the economic value of water and the economic benefits accruing to water has been a key determinant of water reforms leading to better water pricing...also achieving the desired levels of coverage and quality of service requires that the delivery of the services take place under systems that are financially and economically viable.'

Conclusion

Water must be treated as one of Earth's most precious natural resources. It is the driver of all economic and social development and provides invaluable services as an aquatic ecosystem. Given the threats posed by climate variability and human induced climate change, it is anticipated that water will be a critical issue in decades to come. Thus, Integrated Water Resource Management is gradually being given greater consideration by policymakers and water users in Guyana. Current initiatives for institutional reforms, such as the establishment of a National Water Council, demonstrate the Guyana Government's commitment to addressing the critical institutional issues and threats. Nonetheless, it needs to take urgent and decisive action on realistic pricing and the supply of clean water. Further, the provision of reliable data and information will provide the basis for wise decision-making. The way forward is the immediate resuscitation of the National Water Council and the completion of the National Water Policy and Integrated Water Resource Management Plan to create an overall mechanism that is adaptable to environmental changes; along with this, it is also necessary to set definitive water standards. Together, all these measures will collectively provide the basic institutional framework within which wise management decisions and actions can be taken.

References

Bynoe, P. and M. Bynoe (2006). *Report on An Appraisal of the Environmental Impact Assessment Process and Procedures, as well as the Permitting Systems in Guyana.* Georgetown: Inter-American Development Bank.

Bynoe, P. and P. Williams (2007). *Final Report on Guyana Study of Biodiversity Management in the Amazon.* Brasilia: Organization of the Treaty of Amazonia Secretariat.

Bynoe, P. (2008). *Situational Analysis of Watershed Management in Guyana Report.* Prepared as part of a Watershed Analysis and Management Study. Georgetown: Guyana Lands and Survey Commission.

Cummings, A. (2006). *National Vision for Land and Water Use.* Final Report. ACTO/GEF/UNEP/OAS.

Davis, D. (1996). "Water Resources Assessment: The Tool for a Sustainable Future", in *Water Resources Assessment and Management Strategies in Latin America and the Caribbean.* Proceedings of the WMO/IDB Conference. San Jose.

East Demerara Water Conservancy Act. 1998. Chapter 53:03.

Environmental Protection Act, No. 11 of 1996.

Environmental Protection Agency (2000). *Integrated Coastal Zone Management Plan.* Georgetown.

————. (2006). *State of the Environment Report for the Demerara Watershed.* Prepared by SENES Consultants Limited. Georgetown.

Food and Agricultural Organization (2010). *Caribbean Water Policy: Spanning the Spectrum.* Rome.

Government of Guyana (2000). "*Water. National Development Strategy (2001-2010)*", Chapter 15. Georgetown: Ministry of Finance.

————. (2000a). "Water Management and Flood Control Policies", *Chapter 40, National Development Strategy (2001-2010).* Georgetown: Ministry of Finance.

————. (2005). *Poverty Reduction Strategy Paper.* Georgetown: Ministry of Finance.

————. (2006). *The Amerindian Act Guyana.* Georgetown.

————. (2007). *National Biodiversity Action Plan II (2007-2011)*. Georgetown. Environmental Protection Agency.

Guyana Climate Change Action Plan, 2001. Global Environment Facility, UNDP.

Guyana Geology and Mines Commission (1994). *Environmental Mining Agreement.*

Guyana Forestry Commission (2002). *Annual Plan Guidelines for Timber Harvesting.*

————. (2002a). *Annual Plan of Operation Guidelines for Conservation 2002*

————. (2002b). *Code of Practice for Timber Harvesting.*

Guyana Housing and Population Census, 2002.

Inter-American Development Bank (n.d.). *Integrated Water Resources Management: Institutional and Policy Reform.* Environment Division, Sustainable Development Department.

Inter-American Development Bank (1998). *Strategy for Integrated Water Resources Management.* No. ENV-125. Washington. December.

Janki, M. (n.d.). "Customary Water Laws and Practices: Guyana". *http://www.fao.org/ legal/advserv/FAOIUCNcs/Guyana.pdf*

Lanna, A.E. (n.d.). "Water Charges in Brazil: Implementation and Perspectives". Inter-American Development Bank website: *http//wwww.iadb.org* (Downloaded October 12, 2010).

Lee, T. (1997). *Water Resources Management Issues and Challenges in the Caribbean.* Environment and Development Division United Nations Economic Commission for Latin America and the Caribbean Santiago: Inter-American Development Bank.

Mining (Amendment) Regulations Act 2005.

National Action Plan for Combating Land Degradation (2006).

Perret, S., S. Farolfi and R. Hassan (eds.) (2006). *Water Governance for Sustainable Development: Approaches and Lessons from Developing and Transitional Countries.* London, UK: Earthscan.

Rogers, R. (2002). *Water Governance in Latin America and the Caribbean.* Washington: Inter-American Development Bank-Sustainable Development Department. Environment Division.

Saleth, R.M. (2006). "Understanding Water Institutions: Structure, Environment and Change Process", in Secretariat of the United Nations *Convention to Combat Desertification.* Draft UNCCD Water Policy Advocacy Framework for: *Securing Water Resources in Water Scarce Ecosystems in the Face of Desertification, Land Degradation and Drought (DLDD) Imperatives.*

San Martin, O. (2002). *Water Resources in Latin America and the Caribbean: Issues and Options.* Washington: Inter-American Development Bank.

United States Army Corps of Engineers (1998). *Water Resources Assessment of Guyana.*

UNCED (1992). *Guyana's Report to the United Nations Conference on Environment and Development.* June.

USAID (2009). *Addressing Water Challenges in the Developing World-A Framework for Action.* Bureau of Economic Growth, Agriculture and Trade, U.S. Agency for International Development. March.

Water and Sewerage Act (2002).

Section III

Water Policy Perspectives

Contested Constructions of Water in the Policy Arena

New Challenges and Opportunities

SHAILAJA FENNELL

Introduction

Water is a scarce resource that is at the heart of political and social contestation in local, subnational and national arenas in South Asia. The power that water wields is now manifest in both scientific and social science research agendas. The consequences of a drastic reduction in water flow in South Asian rivers is prominent in the earth sciences while the danger of large dams to local ecologies and livelihoods, most famously the case of the 'Narmada Bachao Andolan', has become a global symbol of poor ecological management. The strength that is unleashed by an institutional management of water is also manifestly evident, with historical analysis pointing to state institutions gaining considerable clout from ownership of water, the most extreme being in the form of 'oriental despotism'. And the current phenomenon of the marketisation of water raises concerns about unfair advantage to the private sector and exclusion of the poor.

There are new solutions for water management currently emerging from new interdisciplinary work. In particular, the subfield of institutional design of natural resource management has identified that institutional mechanisms must incorporate both the technical dimensions, such as scale of the resource, as well as social specificity, such as heterogeneity of users, to ensure sustainable solutions (Ostrom, 1990; 2005).

There are now specific formulations for advancing our understanding of water resource management that incorporate the presence of water resources in different physical forms, river and groundwater, as well as the different use by location, rural and urban, to provide finer distinctions on the supply-side features of 'fit'. The 'fitting' of water management to administrative and hydrological boundaries, of ensuring 'interplay' between water management with other forms of governance, and evaluating dimensions of 'scale' to connect water management and administration at various levels have been uncovered in recent institutional analysis (Mollinga, 2008). There has also been more careful analysis of the demand-side institutional differences that emerge in the management of natural resources due to the contextual aspects that are based on different needs in individual countries (Bandaragoda, 2006). The importance of emphasising the need for identification and incorporation of a myriad of supply-side and demand-side features in relationship to water ownership and pattern of use requiring both scientific and social science perspectives has begun to underscore the need for further interdisciplinary framework for sustainable water resource management.

The implications of the new thinking in institutional design point to a need for policy formulation on water management to ensure careful consideration of disciplinary perspectives. There remains a tendency for single discipline-based policy advisers to rather than move towards an interdisciplinary framework of water management. This restrictive policy framing makes it difficult for policymakers to identify the most appropriate academic construction of water to provide the blueprint for policy design and analysis. Moving to an institutional analysis of water that begins by identifying the difficulties that face natural resource management within national policy frameworks where there has been a single disciplinary framing, is a useful starting point for interdisciplinary framing.

This paper uses the case of forest management as illustration of the challenges that natural resource management face and to show how an institutional analysis could improve our understanding. This is followed by a discussion of how a construction of water resource management that draws on a range of social science disciplines can identify gaps in the valuation of water to improve estimation of both returns as well as risks. The chapter concludes that giving a greater social contextualisation to access and ownership of water resources will permit a better balance between demand- and supply-side aspects of water management and bring in a fuller array of stakeholders to ensure sustainable water resource management.

Institutional Theories of Natural Resource Management

Natural resource management has been the subject of study with a number of science and social science disciplines—ranging from botany to management studies—within the academic realm. Within social science thinking, a shift took place in the 1980s, from a singular focus on economic valuation methods using quantitative techniques such as cost-benefit analysis to an understanding of the technical and social features of a particular natural resource.

The move from market economics to institutional design came about as part of a new understanding of how problems of the environment should be addressed. The starting point was admission among economic analysts that there was no longer a possibility of regarding natural resources using a 'one-size fits all' approach as the topographical and social particularities of natural resource, e.g., whether water was located in an inland lake or a river valley, and the nature of the groups with access to the resource, needed to be given far more attention than previously presumed (Ostrom, 1990).

The importance of the physical and social context within which a natural resource is managed provided the starting point for new cost calculations that focussed on how to ensure that individuals conserved resources. The framework of analysing water management that was devised within the new school of New Institutional Economics (NIE) focussed on identifying economic costs associated with devising mechanisms for the conservation of resources. The NIE framing regarded the property rights assignment among individual users of the natural resource as central to ensuring efficient resource use and thereby conserving the natural resource. NIE thinking, consequently, directs governments and other institutions to give primary importance to the pattern of the allocation of ownership, favouring those individual users who have the greater stake in a resource as priority owners (Libecap, 1989; Ostrom, 1990).

The impact of such a framing—that has been influenced by the North American rather than the European approach to institutional analysis—in the social science has resulted in a greater interest in setting out contracts between government and individuals (or groups) rather than in mapping the complex use of the resource by individuals and communities.[1] The preference in the NIE approach for identifying natural resource management with particular owners and users rather than the pattern of use and its relationship to the social context has resulted in a narrow notion of management rather than a socially constituted basis for sustainable resource use.

The limitation of the NIE framework lies in its assumption that optimising individual preferences drives the decisions of individuals in the management of natural resources (Saravanan, 2009). The nature of natural resource management is consequently reduced to the desires of individual players and does not incorporate the larger social context within which these decisions are taken. The focus on individual decisions also prohibits an explicit incorporation of norms that might operate in time periods that exceed that of an individual's life (Fennell, 2009). In particular, it is not helpful for understanding the role of factors that function across the generations and have inter-generational effects.

The contribution of long-term social norms in managing natural resources in the lives of local communities has been disregarded by NIE and also been forgotten by the mainstream of development institutions in the second half of the 20th century. This oversight has resulted in supply-side decisions being made by national governments and their designated institutions regarding ownership and contracting being undertaken without taking on board the demand-side considerations that affect the livelihoods of local communities dependent on these natural resources. The inability to incorporate the role of social groups demanding access to, and management of natural resources has resulted in a partial and lopsided understanding of management design.

The shortcoming of traditional economic thinking and limitations of the NIE thinking on institutional design expose the limits of costing and contracting approaches to natural resource management. The need for policy formulation on water management to go beyond single disciplinary

---

1. The classic NIE position of contracts in natural resources can be found in the path-breaking work of authors such as Libecap (1989). The European approach to understanding the management of natural resources is grounded more in the study of the commons within its historical context (see Hodgson, 1993 on the difference between the two approaches).

perspectives and comprehensively take on board social and technical dimensions has been emphasised by critics of these approaches (Ostrom, 1990). This restrictive policy framing also denies the role of communities and groups in conserving natural resources and prevents them from being brought in as active agents who can contribute to the designing of the most appropriate blueprint for policy design and analysis.

The Social Significance of Natural Resource Management

The significant role played by communities in the management of natural resources has become an important area of study in ecological social sciences in the last decade (Cornwall and Scoones, 2011). The subfield of forest management has been particularly helpful in improving the understanding of social scientists of the challenges posed by state policies that do not take into account the needs and perspectives of local forest communities. These studies focus on the institutional diversity that is present in both the supply- and demand-side features of natural resource management using a combination of natural and social science tools (Ostrom, 2005). The purpose of these new techniques in devising natural resource management system is to ensure that faulty methods drawn from narrow single disciplinary approaches are no longer applied in institutional design.

One subfield of natural resource management where the dangers of narrow approaches to natural resource management became evident by the end of the 20th century was that of forestry (Moran and Ostrom, 2005). The challenge was sharply etched with regard to the diminution in forest cover across the globe, a matter raised by climate change activists most sharply in the case of the Amazonian rainforest. The reduction in forests throws up a number of national contexts where it is evident that supply-side concerns far exceeded the demand-side analyses of the manner in which forests were managed. The evidence in the national examples below shows the consequences of such a one-sided approach to natural resource management.

Lessons from the Management of Asian Forests[2]

The increasing colonial demands for Asian wood created a very important commodity trade during the 19th century in South Asia and East Africa (Sivaramakrishnan, 1995). The lucrative market opportunities arising from the considerable commercial value of particular species of timber resulted in private contracts being provided by the state for the removal and use of wood and other forest produce. While such contracting was initiated during the colonial era, it continued into the post-colonial period. There was little pause for thought, or consideration, for the historical trail of denial of community rights in and usage by the public of, the forests as livelihood and social structure (Guha, 1989). This disregard displayed by the state for traditional, communal and indigeneous rights to the forest resources met with confrontation from the community and groups in civil society.

The clash over the ownership of the forest and its primacy as a form of life and livelihood has been wrought with distributional conflict and the contestation of rights.[3] At the heart of this conflict is the manner in which state power uses politics, rhetoric and knowledge to give the highest priority to current economic value and market opportunity over cultural, traditional and indigeneous forms of society. In the case of forests, there is a need to undertake a full study of the history of ownership and stakeholder usage for 'there is a dialectical relationship between discourses of rules and discourses of protest, and we can advance the study of this relationship by treating resistance as a diagnostic of power' (Sivaramakrishnan, 1995: 3). The nature and reason for contestation in the following forest struggles underlines the problems that emerge when demand-side factors are not adequately recognised and incorporated into policy design.

---

2. This subsection is based on Fennell (2009).

3. The tendency to overlook the rights inherent in the way of life of groups that have traditionally resided and drawn on the forest persists across modern states. The conflicts between tribals/indigeneous forest people and the modern state pepper the history of the 20th century, such as the case of the Chiapas in Mexico and the Orang Asli in Malaysia.

Malaysia

The case of the Orang Asli, the original peoples of Malaysia who were excluded from any legal deed to their traditional lands, sets out the difficulties encountered by local communities getting formal recognition of their systems of natural resource management. The British colonial administration took the formal position that they were supportive of the laws promulgated by the princely Malay states and such laws decreed that land was the property of the kings. After independence, this omission was carried over into the national policies of the Malaysian government. The National Land Code to provide a uniform system of land ownership that was introduced in 1965 continued to draw on the Australian Torrens system of land registration that had initially been brought in by the British colonial administration during the 1930s (Means, 1985).[4] As independent Malaysia was a federation consisting of 13 princely states, the federation required the promulgation of laws that operate across all states. The National Land Code proceeded to vest land rights of individuals only upon registration in the land registry. This was then approved by each of the princely states but the legislation was contradictory to the practices of indigeneous groups. In the case resource use norms in operation among the 12 tribes of the Orang Asli, the practice was to pass on their collective rights from generation to generation through customary law, and not to vest it in an individual (Cheah, 2004).

The inability of state law to recognise the rights of the Orang Asli became a matter of public attention when the federal government used the Land Code to compulsorily requisition the land of the Orang Asli Temuan tribe, to facilitate a road link between the new Kuala Lumpur International Airport and Kuala Lumpur city. The Temuans took their complaint to the High Court of Malaysia and sought the restitution of their right to ancestral land. In its 2002 decision, the court recognised that the rights of the Orang Asli were different from the private land rights determined within market contexts. The court recognised that the Orang Asli did not obtain only livelihoods, i.e., economic products from the land, but also that their very way of life was directed by the spirits of their ancestral land. It therefore decreed that the 'native title' of the Orang Asli could not be

---

4. The Torrens system of land ownership is based on a Hanseatic system of land registration that was initially introduced into Australia by the British colonial administration.

treated as though land was a mere commodity and on par with private land-holdings, rather it should be regarded as a way of life that was based on a system of beliefs linked to the land (Cheah, 2004).

India

The colonial policies of forest management that had been brought in during the 19th century did not undergo changes till the 1970s. Forest agitations, most particularly that of the Chipko agitation in the mountainous regions of Uttar Pradesh, resulted in changes in the forest legislations in the 1980s (Damodaran and Engel, 2003). The wresting of rights for local hill communities from the government is laudable, yet there continued to be difficulties in ensuring that subsequent forest policies did incorporate the demands for acceptance of traditional ways of resource access that were made by the local community.

The difficulties of getting a balance between state policy and community demands was attempted through a new form of resource management termed as Joint Forestry Management (JFM), introduced in the 1990s. The JFM was designed to bring about a participatory form of forest management. However, these JFM programmes failed on account of official interference and were unable to vest any power in the local communities, whether by intent or poor design, and hence the district initiatives remained short-term and enabled premature exploitation of the trees (Sunder, 2000).

This new managerial form of organised resource use continued to be the official state framework for a wave of social forestry programmes in Indian districts for the next two decades. The participation that was to be the foundation of this framework did not result in the poorest sections of the forest community, by caste or occupation, gaining equal access to forest resources (Vemuri, 2008). The failure of the management programme to restructure the access structure indicates that the social features of demand were not taken into account adequately in what remained a top-down management structure of the bureaucratic administrative and forest services.

China

The difficulties faced by the state in recognising and designing common property resources are widespread and cross over different development paradigms. On independence in 1949, all land, including forest land, in China was declared to be state property. The Chinese approach to forest the management during the first half of the 1950s was modelled closely on Stalin's understanding of controlling and exploiting nature (Bao, 2006). In the period from 1956 onwards, forest land was awarded to private households but this policy was revoked during the years of the cultural revolution from 1966-1976 (Long and Zhou, 2001). It was only after the start of the economic reforms of the 1980s that there was a reconsideration of the rights to the forests with regard to usage and management of resources.

On September 20, 1984, the National People's Congress Standing Committee adopted the Forest Law of the People's Republic of China. This law was formulated to protect, nurture, and rationally utilise the forest resources so as to speed up the greening of the country's territory. The law was designed to bring into prominence the roles that the forest could play with regard to storing water, saving soil, adjusting the climate, improving the environment, and supplying forest products (Xi, 1999). The remit of the law was largely with regard to changing incentives with regard to the conduct of forest and forest tree cultivating, planting, logging and utilisation, and to regulating the operation and management of forests, trees and woodlands.[5] The strongly developmental spin placed on the use and conservation of forest lands has implications for the rights and lives of indigeneous people who have long established customary laws with regard to the forest lands in far-flung areas of the country. The sale of user rights and the growth of tourism have emerged as large revenue earners for the provincial governments but they have also come into conflict with the traditional ways of the indigeneous groups such as the Dai and Junuo in the province of Yunan. The market-based approaches to user rights have devalued traditional ways of knowledge transfer regarding flora and fauna (Liu, 2007).

The politics of forest resource management is contest-ridden in Asia. Indigeneous and tribal groups continue to be overlooked by the bureaucratic and forest institutions and supply-side matters such as contracts take the upper hand. Where laws do exist, they are rarely invoked by the local communities who prefer to turn to customary laws to deal with disputes in

---

5. The Forest Law was amended on April 29, 1998, to introduce the transferability of use rights in a range of forest types: timber forests, economic forests, fuelwood forests, so that these could last up to 70 years and be renewable.

the forests. The gap between contractual, supply-side models of forest use and the group demands arising from social norms of forest use indicate that all is not well in the world of forest policy modelling.

There has been an attempt to move away from a divisive confrontation between state and civil society towards community approaches that use interdisciplinary tools to ensure sustainable forest resource management. The biggest challenge to managing these resources is the difficulty that players on the supply and demand sides face in working towards creation of a common platform through a process of 'collaboration as a way forward' (Vira *et al.*, 1998).

This collaborative process suggested by the heterodox multi-disciplinary thinking is based on the understanding that where all players have a shared urgency of risk in natural resource use there is more likely to be sustainable resource management. The central plank in this approach to natural resource management is that if all stakeholders identify a common need to use a valued resource, it will permit the use of procedures to reduce risk by using collective mechanisms to mitigate costs. If such collective mechanisms can be made to work, despite the presence of social and economic hierarchies, the process of collaboration is likely to succeed. In contrary situations, such as those seen in the development trajectories of the forest sector in Malaysia, India and China, there has not been adequate effort to identify all the players who are stakeholders on the demand and supply side. Consequently, the institutional design for resource management in such situations will suffer from poor computation resulting in an underestimation of relative costs and benefits (Adams *et al.*, 2002).

The lessons from the poor record of forest resource management relate to the difficulty of identifying the full range of players on both supply and demand sides of natural resource management and usage. Simple interventions based on legislative reform do not provide a sufficient basis for natural resource management as the legal framework alone is not able to ensure equality of access for all players nor can it reverse top-down bureaucratic administrative and forest institutional structures.

The limitations of these early management frameworks have led to a review of natural resource management that goes beyond contracting and formal laws to a more careful inclusion of social norms and informal practices (Ostrom, 2005). The importance of carefully linking law to policy and administration has also been recognised with regard to water resource management. In particular, water law, policy and administration have been identified as three pillars of water management in international literature (Saleth and Dinar, 2000). These formulations are also finding favourable reception among scholars of Indian water policy.

Institutional Management of India's Water Resources

The focus on law, policy and organisations as central themes for an improved institutional analysis of water management has been welcomed but there is concern that these global policy formulations have yet to devise tools that can analyse how societies adapt to the supply-side interventions undertaken by the government and other institutional players (Shah, 2005). Additionally, there are scholars who are concerned that the lessons from the international sphere should not be seen as the single mantra that can solve water management problems. In fact, the reverse process by which Indian water resources experiences have also fashioned national and international water policy frameworks in past decades should also be recognised (Mollinga, 2010). The possible solutions for Indian water resource management lie in the overlapping spheres of these two processes. The complexities of India's water bodies require detailed knowledge of social and technological processes (Shah, 2003) alongside a more astute evaluation of the hydrological and other ecological dimensions (Bandaragoda, 2006).

The case of rural water management has been of particular significance in devising management structures. The context of this phenomenon emerged with the rising power of the rural farmers' lobbies as rural capitalism began to emerge in India in the 1960s (Brass, 1995). Another dimension of rural action emerged with the increasingly vocal protests by subordinate groups such as marginal farmers and landless labour in anti-dam movements such as the 'Narmada Bachao Andolan' from the 1980s (Baviskar, 1996).

The political economy of rural water management has been played out in the battleground between national architects of development policy and local advocacy groups for the poor and marginalised groups in the last few decades. The backdrop to this story of conflicting claims to water has been provided by the irrigation department, which has grown in importance with the increasing demand for water for the thirsty high-yielding varieties (HYVs) brought in by the Green Revolution technologies of the 1960s. The associated increase in rent-seeking opportunities alongside the growing revenue accruing from water delivery became a subject for national and international academic analysis (Wade, 1985; Mosse, 2003). The political posturing of the rural elites and their use of irrigation facilities to gain private wealth through the proliferation of corrupt practices in water delivery and usage became endemic in the sector. It is noteworthy that it is the unpacking of the social context of water management that revealed specific characteristics and types of water resource management in the case of rural India.

The technical aspect of such inequalities has also received attention. The first feature here is known as the 'head end—tail end problematic'. This terminology relates to the topographical feature in an irrigation system where farmers on the upstream side of a canal are able to appropriate more than their share of the water, those at the downstream side being deprived. The resource consequences of this are that there is an inequity in water access and usage. Such differentiation is superimposed by the fact that richer farmers tend to occupy or to manipulate access to the upstream land and this results in an unhealthy overlay between social and spatial inequality with regard to rural water resources (Mollinga, 2010). The consequence for Indian rural water management has been that there has been no strong lobby to demand irrigation reforms and where there have been activist movements, the matter of water distribution has not come to the forefront in any consistent manner.

The second aspect of technical differences in irrigation relate to the heterogeneity of irrigation types: for instance, how badly irrigation has been overlooked in relation to canal irrigation. In the former, water is restricted to examination of a single source while in the latter case, water availability is dependent on the ability of individual tube well pumps to draw up groundwater. The preference in the economic approach to water management to focus on the existence and operation of water markets is related to the importance of tube well-extracted water for the accumulation of rural wealth in regions such as the Punjab. It is the economic costs rather than the social and political aspects of water access to non tube well owners and inequities that exist with regard to small farmers and tenants that have been the subject of study (Dubash, 2002). The existing management structures for groundwater extraction have focussed on a contractual framework that is based on rationing of water extraction across claimants. Such an approach is ineffective as it does not permit an institutional analysis of individual household demand or the pattern and purposes of water use by each household (Shah, 2005).

The third aspect of the technical features of water resource management is that of false conceptualisation as in the case of tank-based irrigation. Despite the tank being a human construction that is located within a community, there is still limited analysis of the social differentiation in a village and this has resulted in poor management practices in tanks. This shortcoming arises out of a shallow understanding of community and a hastiness to regard small villages as non-hierarchical spaces. This is a manifestly faulty proposition as there are caste-based and use-based conflicts in water usage that are prominent in these communities (Shah, 2003). Tank irrigation is therefore presented as a simple form of water resource use and conservation. This is not accurate for there are social features and social relations that affect water usage: e.g., where elites undertake mechanical recharging of water basins that change the water levels in tanks in the locality and consequently the water availability to households from traditional water use arrangements.

There are numerous shortcomings of a narrow economic and technical analysis of Indian water resources that has been highlighted by Indian scholars. Furthermore, this failure is compounded by the existing management practices undertaken by administrative and water bureaucracies at district, block and cluster levels in rural India. These bodies continue to operate within a top-down institutional framework to impart justice in rural areas. This has resulted in a history of community confrontation against the state to wrest away group rights to natural resource ownership.

The claims by subordinate and marginal groups have been countered by national development architects who advocate a trickle-down effect that will benefit these groups as economic growth proceeds. This market-based approach to development has been unable to accord rights to marginal farmers, landless labour and other marginal groups in local communities and their rights to water resources have consequently diminished over the last few decades. This situation prevails despite the attempt of new management approaches to include the 'community' while the water resources that are located in these communities are becoming attractive to the state as contributors to new sources of wealth (Shah, 2003). This head-on collision results in conflicts regarding the institutional design of natural resource management frameworks between the supply-side interventions being devised by bureaucrats, national and international agencies and the patterns and purposes of use among heterogeneous users of water across a range of water resource types (Shah, 2005).

The inequity perpetuated by such poorly designed institutional mechanisms for water management has resulted in regressive social and economic consequences within local communities. Furthermore, there are additional social fractures caused by the overlay of social and spatial differentiation that is present in the management of water resources in India. These complex social features imply that legal reform cannot ensure that water resource management can be improved through remedying ownership assignment. The reason for this shortcoming is that legal reform in a narrow economic framing will be restricted to those who have rights of ownership, such as landowners, tube well owners and tank irrigation owners, but regarded all other groups as tenants of the state (Fennell, 2010). This binary separation of members of a local community creates a basis for conflict within the community that works against forging a common identity that is necessary for sustainable natural resource management.

The focus on identifying ownership and contractual rights does not take into account the processes or purposes of usage by those, in the community, who do not have access to these privileges. So legal reforms that reorganise the principles of rationing water, or even redistributing water use, will only value the economic benefit from such water usage but not take into account livelihood or cultural dimensions that do not have a market equivalent. A legal perspective that reduces individuals' varied use of water to mere economic motivation and eschews the cultural, political and social experience of water usage results in poor computation of the costs and benefits of particular mechanisms of water resource management. It also militates against a securing of all three pillars of water management; law, policy and administration, required for sustainable institutional design.

If the institutional design of water resource management mechanisms does not take into account the potential power of bringing in tenets of social justice in relation to both supply- and demand-side features, there is less chance of a common identity around resource use. Social justice defined in terms of equity rather than individual equality appears to be a more effective starting point as it would permit that individuals be treated 'fairly' so as to address social and economic inequalities. The call for equality based on the notion that the starting point for all individuals are identical and can therefore be fully functional in a market context without caste, class or gender hierarchies is not likely in a case where there are a variety of inequities present in water access and use (Fennell, 2010).

Reframing the Management Institutions for Water Resources

The limitations of existing water resource management mechanisms in India cannot be rectified by looking only to legal reform for solutions. Furthermore, there should be measures to ensure that legal changes are required to dovetail with policy and administrative reforms. This requirement is particularly pertinent in the current legal environment where the right to property is regarded as sacrosanct. It was only after the green revolution in agriculture became widespread in India and the powerful landlords became rural capitalists that the state changed its views on private property in landholdings (Sathe, 2003). These forms of persistent inequity in treatment by the courts have cast doubt on the ability of the law to transform social relations.

It is the current obstacles to ensuring that legal reform can be transferred to policy and administrative spheres that make the new inter-disciplinary mechanisms based on multidisciplinary perspectives a promising new approach. The possibility of using identifying institutional features such as the 'fit' between administration and hydrology provide a way to disentangle parts of supply features that could make for more careful legal redress such as regulatory reform of delivery mechanisms and the monitoring of use; in the case of 'interplay' there could be different methods of creating synergy between water management and other local management structures—whether these are social or market-based could be determined by the use of participatory evaluation; and finally 'scale' could be the incentives provided for both administrative and water bureaucracy to reduce any rent-seeking opportunities from top-down management.[6]

The possibility of moving from across the three pillars from administration to policy and finally the law permits the creation of community-led initiatives that focus on the revealing of both the process and the purpose of water usage by each household in the community. Using the water and bureaucratic administration methods to identify household usage can provide contextualised and location-specific evidence on water availability, access and ownership. The heterogeneity, and possible conflict, between diverse groups in the community can be an entry point for uncovering the challenges to establishing a platform for resource management. In recent years, there have been academic initiatives to use multidisciplinary teams and devise interdisciplinary approaches to devise a broader basis for conceptualising the management of natural resources that takes into account the social, political and ecological systems that interact with the economic motivation of individuals (Poteete *et al.*, 2010).

The identification of the specific social, cultural, political and economic values that each group accorded to water availability and access can also contribute to better calculation of the benefits and costs of current ownership patterns of water resources. For instance, if there is evidence that the lives and livelihoods of marginal groups are most adversely affected by existing water ownership patterns, it would be a more effective strategy to provide new water markets that favour the most disadvantaged, whether this is marked by case, class or gender. It is in the possibility of using the contextual and specific features of differentiation that new opportunities for water markets and values can be ascertained for use in new interdisciplinary tools for constructing management frameworks. The ability to bring in social and power relations through the mapping of lived experiences is a these novel frameworks facilitate the study of individual

---

6. See Mollinga (2010) for a useful discussion of each of these aspects.

choice is terms of community norms and locally constructed attributes and values (Cornwall and Scoones, 2011).

The move from earlier governmental and market notions of ownership of natural resources towards models that draw on stakeholder management models facilitate collaboration across groups in maintaining the resources in the water sector. Simultaneously, the movement away from legal diktat for natural resource management frameworks to an administrative-based process that permits a process and purpose-based mapping on heterogeneous demands within a local community ensures that both demand and supply features of institutional design can be tested and subsequently monitored when the framework is operational.

Conclusion

The growing concerns about water, both globally and within India, have brought to the fore the difficulties of creating a platform for sustainable water resource management when there is a policy environment of conflict. The divisive elements are not only in the actual field of water management but also in the combative attitude taken by individual disciplines in the social sciences regarding the appropriate policy framing for the institutional design of water resource management. This provides a challenge to policymakers in identifying the most appropriate academic construction of water to provide the blueprint for policy design and analysis.

The alternative framing of academic disciplines in new resource management analysis show how the intersections of various disciplinary perspectives can contribute to a more nuanced understanding of the supply- and demand-side features of water resource provision and usage. The most important feature is the ability to bring in contextualised, and location-specific analysis of water availability, access and ownership. The possibility of using these process and purpose-based mapping of heterogeneous usage within a community provides the beginning of a water resource design that has the ability to identify the difficulties of attempting natural resource management within national policy frameworks. Such a framework permits a construction of water drawing on natural, scientific and social science disciplines to identify gaps in the valuation of water, leading to poor estimation of both returns as well as risks, in a range of environments.

As was indicated in the review of challenges of forest resource management, it is clear that following a narrow approach that focussed on economic valuation and regarded law as the starting point for devising a framework for natural resource management is fundamentally flawed. The need to regard policy as primarily constituted by political and social contestation to prevent further reduction in the water resource availability in India points to legal reform as the final stage rather than the starting point for institutional design. Undertaking an analysis of state institutions and how they regard ownership of water, particularly in relationship to ownership and contractual considerations, shows that there is an exclusion of the poor and marginalised groups in the local community.

Focussing on governance mechanisms for sustaining natural resources that are based on the particularities of water resource type management mechanisms to take into account both the technical dimensions can create better fit, interplay and scale dimensions within any management mechanism. In the case of water resources, its presence in different physical forms, canal, tube well and tank, as well as the particularities of the rural political economy in India, shows that there appears to be a more effective set of opportunities for sustainable water resource management by focussing on those individuals and households most inequitably treated by existing ownership assignment. This use of equitable rules of inclusion can remedy existing binary forms of social structure with regard to resource use and facilitate the creation of a common platform that is required for sustainable water resource management principles to operate.

References

Adams, Bill, Dan Brockington, Jane Dyson and Bhaskar Vira (2002). "Analytical Framework for Dialogue on Common Property Resources", Lead Paper for *Round Table on Promoting Policy Dialogue for Common Pool Resource Management*. 9th Biennial Conference of The International Association for the Study of Common Property, Victoria Falls, Zimbabwe. June 17-21.

Bandaragoda, D.J. (2006). "Status of Institutional Reforms for Integrated Water Resources Management in Asia: Indications from Policy Reviews in Five Countries", *Working Paper* 108. Colombo, Sri Lanka: International Water Management Institute (IWMI).

Bao, M. (2006). "The Evolution of Environmental Policy and its Impact in the People's Republic of China", *Conservation and Society* 4(1): 36-54.

Baviskar, A. (1996). *In the Belly of the River: Tribal Conflicts over Development in the Narmada Valley*. Delhi: Oxford University Press.

Brass. T. (1995). *New Farmers' Movements in India*. London: Routledge.

Cheah, W.L. (2004). "Sagong Tasi and Orang Asli Land Rights in Malaysia: Victory, Milestone of False Start?", *Law, Social Justice and Global Development* 2. *http:// www/go.warwick.ac.uk/elj/lgd/2004_2/cheah*

Cornwall, A. and I. Scoones (2011). *Revolutionizing Development: Reflections on the Work of Robert Chambers*. London: Earthscan.

Damodaran, Appukuttannair and Stefanie Engel (2003). "Joint Forest Management in India: Assessment of Performance and Evaluation of Impacts", *ZEF Discussion Papers on Development Policy*, Paper No. 77. Bonn.

Davis, J. (2004). "Corruption in Public Service Delivery: Experience from South Asia's Water and Sanitation Sector", *World Development* 32.

Dubash, N. (2002). *Tubewell Capitalism: Groundwater Development and Agrarian Change in Gujarat*. New Delhi: Oxford University Press.

Fennell, S. (2009). *Rules, Rubrics and Riches: The Relationship between Law, Institutions and International Development*. Routledge. Chapter 3.

Guha, R. (1989). *The Unquiet Woods: Ecological Change and Peasant Resistance in the Himalaya*. University of California, Berkeley Press and Oxford University Press.

Hodgson, G. (1993). *Economics and Evolution: Bring Life Back into Economics*. Cambridge: Polity Press.

Libecap, G. (1989). *Contracting for Property Rights*. New York: Cambridge University Press.

Liu, J. (2007). "Contextualizing Forestry Discourses and Normative Frameworks towards Sustainable Forest Management in Contemporary China", *International Forestry Review* 9(2): 653-60.

Long, Chun-Lin and Yilan Zhou (2001). "Indigenous Community Forest Management in Jinuo People's Swidden Agroecosystems in South West China", *Biodiversity and Conservation* 10: 753-67.

Means, R. (1985). "The Orang Asli: Aboriginal Policies in Malaysia", *Pacific Affairs* 58(4): 637-52.

Mechler, R., S. Hochrainer, D. Kull, S. Chopde, P. Singh, S. Wajih and The Risk to Resilience Study Team (2008). "Uttar Pradesh Drought Cost-Benefit Analysis, From Risk to Resilience", *Working Paper* No. 5. M. Moench, E. Caspari and A. Pokhrel (eds.). ISET, ISET-Nepal and ProVention, Kathmandu, Nepal. p.32.

Mollinga, Peter P. (2008). "The Water Resources Policy Process in India: Centralisation, Polarisation and New Demands on Governance", in Vishwa Ballabh (ed.), *Governance of Water: Institutional Alternatives and Political Economy*. pp.339-70. New Delhi: Sage.

————. (2010). "The Material Conditions of a Polarised Discourse: Clamours and Silences of Critical Agricultural Water Use in India", *Journal of Agrarian Change* 10(4): 414-36.

Moran, E. and E. Ostrom (2005). *Seeing the Forest and the Trees: Human-Environment Interactions in Forest Systems*. MIT Press.

Mosse, D. (2003). *The Rule of Water. Statecraft, Ecology, and Collective Action in South India*. New Delhi: Oxford University Press.

Ostrom, E. (1990). *Governing the Commons: The Evolution of Institutions for Collective Action*. Cambridge, New York: Cambridge University Press.

————. (2005). *Understanding Institutional Diversity*. Princeton: Princeton University Press and Oxford.

Poteete, A.R., M. Janssen and E. Ostrom (2010). *Working Together: Collective Action and the Commons, and Multiple Methods in Practice*. Princeton, N.J.: Princeton University Press.

Saravanan, V.S. (2009). "Decentralisation and Water Resources Management in the Indian Himalayas: The Contribution of the New Institutional Theories", *Conservation and Society* 7 (3): 176-91.

Saleth, R.M. and A. Dinar (2000). "Institutional Changes in Global Water Sector: Trends, Patterns, and Implications", *Water Policy* 2(3): 175-99.

Sathe, S.P. (2003). *Judicial Activism in India*. Second Edition. New Delhi: Oxford University Press.

Shah, E. (2003). "Social Designs: Tank Irrigation Technology and Agrarian Transformation in Karnataka, South India", *Wageningen University Water Resources Series*. New Delhi: Orient Longman.

Shah, T. (2005). "The New Institutional Economics of India's Water Policy, 2005", Paper presented at international workshop on *African Water Laws: Plural Legislative Frameworks for Rural Water Management in Africa*. January 26-28th. Johannesburg, South Africa.

Sivaramakrishnan, K. (1995). "Colonialism and Forestry in India", *Comparative Studies in Society and History* 37(1): 3-40.

Sunder, N. (2000). "Unpacking the Joint in Joint Forest Management", *Development and Change* 31: 255-79.

Vemuri, A. (2008). "Joint Forest Management in India: An Unavoidable and Conflictual Common Property Regime in Natural Resource Management", *Journal of Development and Social Transformation* 5. November.

Vira, B., O. Dubois, S.E. Daniels and G.B. Walker (1998). "Institutional Pluralism in Forestry: Considerations of Analytical and Operational Tools", *Unasylva* 49(194): 35-42.

Wade, R. (1985). "The Market for Public Office: Why the Indian State is not Better at Development", *World Development* 13(4): 467-97.

Xi, W. (1999). *Forest Policy, Law and Local Participation in Forest Management in China*. *http://www.iges.or.jp/en/fc/phase1/ir99/1-3-Wang%20.pdf*

10

Water Use Efficiency in the Indian Context

A Regulatory and Institutional Mapping

SHAWAHIQ SIDDIQUI

Introduction

Water in India is owned by the Indian state by virtue of absolute rights provided under the law of the land. This absolute right over all natural water in the country vests the state with absolute power over the life of all its citizens and their economic and domestic affairs and flora and fauna as water is vital for sustenance of life and one who owns water or has absolute rights over it, therefore has absolute power over the life of others.[1] In the economics of scarcity, the sovereign state has the responsibility as well as the right to allocate and utilise water resources in the most efficient way and do justice in the distribution of water. This is based on the assumption that the state will in fact be able to bring about the most efficient use as it is equipped with adequate resources, scientific knowledge, administrative system at the decentralised level, regulatory system and institutional structure to do justice in water distribution and allocation. This assumption does not hold much water in the Indian context.

Experience on water governance in India in the past 50 years suggests that inappropriate forest, irrigation and industrial policies by the government have resulted in massive depletion of water resources and has caused great inequities in the distribution of water. As a result, thousands of irrigation canals and dams have been built in the entire landscape of the country and millions of tanks, common village ponds, *johads* and wells have dried up resulting in the complete change in the prioritisation of water

---

1. Singh (1992).

use and availability of surface and groundwater. Therefore in the present scenario of increasing ecological crisis due to climate change, the whole question of laws relating to water, its allocation, equitable distribution and institutional mechanisms involved in all aspects of water resource management needs to be fundamentally examined.

Access to drinking water has been made a Fundamental Right in India.[2] However, availability of sufficient and safe drinking water remains a challenge. Since the turn of the century, greater focus is being given to this issue due to the phenomena of climate change, the alarming effects of which is being felt all over the world. In India, studies reveal that water resources are most vulnerable to the effects of climate change,[3] given that its impact is more than likely to result in unpredictable shifts in the hydrological cycle, critically affecting the overall water scenario. As a result, different water intensive sectors (besides potable water) such as agriculture and industry are likely to face increasing challenges to meet even its basic water requirement. Common water resources such as village ponds, tanks, reservoirs, *johads*, community wells etc., are also likely to be affected due to the change in the hydrological cycle. In this scenario, it is necessary for the government to take appropriate measures for introducing necessary interventions in order that the available water resources are utilised in an efficient manner.

Further, in the wake of the alarming increase in demand for water and enormity of the prospect of water scarcity, water resource management needs a fresh perspective from the water efficiency point of view. The water efficiency perspective, which is a felt need now, is based on the view that averting the looming global and domestic water crisis requires meeting two great challenges: improving water use efficiency to ensure that the quantity of water reserves remain adequate for humanity's many competing demands and reducing pollution and other threats to water quality, to ensure that the water supplies continue to be usable. It is this perspective on efficient use of water which is the subject matter of close scrutiny and examination from the regulatory, institutional and implementation point of view in India.

---

2. Subhash Kumar *versus* State of Bihar, AIR 1991 SC 420.
3. See study by Ministry of Environment and Forests (2009).

Water Law Framework and Recent Attempts for its Reconstruction

Before we examine the existing regulatory and institutional framework and its preparedness to ensure efficient use of water, it will be useful to glance at the current water law framework in India and the current efforts made for improving water governance. So far, water in India has been regulated under different state laws, predominantly as per number of irrigation Acts that belong to colonial era[4] and through institutional mechanism of public health departments for supply of water to municipal areas. Post-independence, the position on water governance did not change much as the Constitutional design vested state governments with the power to regulate all natural water resources and with the absolute rights to control all water resources. Groundwater had remained immune from absolute state control due to strong property and easementary rights regime. Given this scenario, there is no single legislation at the national level addressing all aspects of water as a single unified resource unlike other natural resources such as forest or biodiversity which have national legislation in place addressing and treating them as a unified resource. States have also done very little on bringing a comprehensive legislation in place for addressing all aspects of water resource management so as to meet the competing water demands in all water intensive sectors. However, the *Twelfth Five Year Plan Approach Paper* by the Planning Commission,[5] GoI, mentions a number of welcoming steps that the Central government intends taking for addressing legal reforms required in the water sector. The paper reflects that the Government of India is very keen to bring a Central water law. This seems to be different from the Central groundwater legislation which has been under active debates and discussions since 1995 and has not seen the light of day till date. Fresh approaches for participatory irrigation management and public-private community partnerships (PPCP) are also being explored.[6] There is concern for water sector reforms at the state level as well. For example, the state government of Rajasthan has taken the initiative to draft a comprehensive water law for the state that discusses primary, secondary

---

4. See Madhya Pradesh Irrigation Act, 1931.

5. *www.planningcommission.nic.in*

6. See "Report of the Working Group on Water", *Approach Paper for the 12th Five Year Plan (2012-2017)*.

and tertiary uses of water and provides for equitable distribution of water to different water intensive sectors.[7] The state of Maharashtra has come up with a new legislation, the Maharashtra Water Resources Regulatory Act, 2005, that establishes a Regulatory Authority for the water management in the state. Several states are also in the process of formulating their groundwater bill. For example, the state of Uttar Pradesh has drafted the Uttar Pradesh Water Regulatory Commission Act that touches upon the regulation of groundwater.

The National Action Plan for Climate Change: A New Gateway for Sectoral Reforms in Natural Resource Laws

In view of the global agreements and response to the increasing risks due to climate change, the Government of India took a rather unique initiative by bringing an Action Plan that tries to cover many aspects of natural resource governance and provides emphasis on sustainable developmental pathways aimed at both adaptation to and mitigation of climate risks. The Prime Minister's National Action Plan on Climate Change (NAPCC) launched in 2008 with its eight dedicated missions strengthens the need for bringing reforms in natural resource laws and natural resource governance in the country by suggesting different strategies and approaches for the improvement of laws and governance and institutional structures that will help India in its climate preparedness. Another aspect of the NAPCC is that it is more accountable to India's international obligations under the United Nation's Framework Convention on Climate Change (UNFCCC) and is not directly linked to the domestic legal framework. These missions are: (i) the Jawaharlal Nehru National Solar Mission, (ii) the National Mission for Enhanced Energy Efficiency, (iii) the Green India Mission, (iv) the National Mission on Sustainable Habitat, (v) the National Agriculture Mission, (vi) the National Water Mission, (vii) the National Mission for Himalayan Ecosystem and (viii) the National Mission for Strategic Knowledge on Climate Change. Each of these missions is a vision document of the GoI and is based on a number of assumptions and contingent events for future action plans. The institutional mechanisms to supervise and implement the mission comprises a Mission Steering

---

7. The Author is involved in making contributions to the Rajasthan Water Law (Draft).

Committee consisting of various ministers from relevant ministries. The Steering Committee is headed by the Prime Minister. Each relevant ministry is the nodal institution for implementing the mission. Thus, for example, the Ministry of New and Renewable Energy is the nodal ministry for implementing the National Solar Mission, the Ministry of Environment and Forests is the nodal ministry for implementing the Green India Mission and so on and so forth. It's worth mentioning that these missions do not have coherent mechanisms to coordinate with each other or provide mutual support for their effective implementation, which is altogether a separate area of study. Here we are concerned with water use efficiency and legal and institutional preparedness to bring the same for which the National Water Mission provides an encouraging start. Thus, the National Water Mission (NWM) becomes a subject matter of our inquiry for understanding the nature of opportunity and support that it creates to aid the objective of introducing water use efficiency and the legal and institutional support required to achieve the objectives under the NAPCC's Water Mission which is a response to India's obligations under UNFCCC.

Water Use Efficiency under the National Water Mission

The National Water Mission (NWM) envisages a number of strategies for integrated water resource management and provides for enhancing water use efficiency (WUE) as one of its key objectives. In the context of this emphasis on WUE under the NWM, it would be important to look at the conceptual notion of water use efficiency under the Mission document and within the existing water law framework.

The National Water Mission envisages achieving 20 per cent enhanced efficient use of water by 2017 by adopting conservation and use-based mechanisms under different strategies. MoWR is currently working towards creating a regulatory and institutional mechanism at the Central level to enhance sustainability of water as a resource and WUE in all sectors by formulating norms for efficient water use. It is henceforth essential that the current strength of the legal framework across all sectors that are responsible for water management and water supply could support WUE be examined and improved if the need is felt to do so. Any inquiry to formulate a new regulatory and institutional structure would be the result of findings of the legal position as it exists today.

Constitutional Position

The Constitution provides for the allocation of subject matters that are to be regulated by the Central and state governments. Under the present Constitutional Scheme, water is a state subject wherein the state government has control over all natural water resources available in a particular state and its supply and management in the respective state.[8] The Central government enjoys powers with respect to the interstate river disputes.[9] The Constitution of India does provide for general provisions with regard to the sustainable and efficient uses of available natural resources so as to sub-serve the common good.[10] There are no specific provisions in the Constitution that mandate efficient use of resources in the technical sense. However, several provisions of the Constitution, importantly Article 21, have been used by the Supreme Court of India in various noted judgments for emphasising the need for sustainable utilisation of country's available resources. The Supreme Court has also emphasised the need for implementing principles of international environment law such as the principle of intergenerational equity, precautionary principle and polluter pays principle. Given this framework, the tough task of introducing efficiency in water use across all sectors does not find specific mandate under the Constitution.

Legal and Institutional Position: National Perspective

The Water (Prevention and Control of Pollution) Act, 1974 is a national law that addresses the issues related to water pollution and contamination. The Act provides for the establishment of national and state-level institutional machinery for the prevention and control of

---

8. Under the Constitution of India, 1950, water is a state subject. The Central government intervenes in the interstate water disputes. The legislative power to make laws in our polity related to water, that is to say, water supplies, irrigation and canals, drainage and embankments, water storage and water power, has been granted to the states by our Constitution. This power is subject to the powers of the Central government to regulate and develop interstate rivers to the extent declared by Parliament. The Central government also has the power to make laws for the adjudication of any dispute relating to waters of interstate rivers or river valleys under Article 262 of the Constitution. However, it would be pertinent to mention here that about 90 per cent of the territory of India is drained through interstate river basins.

9. Article 262, Constitution of India, 1950.

10. Directive Principles of State Policy under the Constitution.

pollution. This is the only water related legal instrument at the national level brought forth under Article 252 of the Constitution of India whereby it is lawful for Parliament to enact a law not mentioned in the Union List under Article 249 and 250 (Union List) at the request of two or more states. However, this central law addresses only the pollution aspect of water and does not address water-related issues considering water as a unified single resource and its efficient use leading to enhanced conservation and sustainable utilisation of available water resources. Another important legislation at the Central level, the Environment (Protection) Act, 1986 (EPA), provides for a very strong mechanism for the protection of the environment. The EPA empowers the Central government to take any appropriate measures that it deems fit for the protection of the environment and restoring its wholesomeness. In the instance where water use efficiency is to be introduced as a legal norm with institutional support to implement it under the EPA, the specific provision of the EPA Section 3(3) assumes great significance. This provision has been widely used by the Central government for the creation of various authorities for the protection of the environment, including the Taj Trapezium Authority, the Dahanu Taluka Authority, the National and State Coastal Zone Management Authorities and the National Environment Appellate Authority (that has been replaced with the National Green Tribunal in the recent past). However, EPA does not specifically provide for a mechanism for introducing efficiency norms for the efficient utilisation of finite natural resources. What it certainly provides for is the broad mandate to constitute an institutional mechanism for introducing and implementing WUE. Given this broad framework and general nature of the EPA, the specificity and focus of an issue such as WUE gets cornered. Perhaps a more focussed regulatory and institutional mandate is required for improving efficiency in the water use across all sectors.

Other state-specific water laws including the recent enactments that are progressive in nature such as the Maharashtra Water Regulatory Act and Uttar Pradesh Water Regulatory Commission Act do not address the issue of efficiency in water use.

There are other subject matter laws that become indirectly relevant for understanding the issue of efficiency from a sectoral perspective, such as irrigation laws and laws for industrial siting and water resource allocation;

these legislations and the critical role played by institutions such as state industrial corporations could not be specifically discussed due to the enormity of these legislations and the complex nature of key subject matter that is under examination here. However, the current status of laws in addressing WUE and key barriers that would be required to be removed for implementing WUE emerge from the study of the main water law framework discussed above (and its limitations discussed below, identified as one of the barriers to WUE) and sectoral laws as summarised below.

The Current Water Law Framework does not Address the Issue of Water Use Efficiency

As has been discussed above, the Constitutional scheme to water resource management provides all powers in the hands of the state government for managing water. At the same time, the Constitution provides enough scope for taking measures for the efficient utilisation of these vital resources, so that it subserves the common good.[11] There are other important provisions in the form of Article 48 A and Article 51 (g) that provides for the state's responsibility for the protection of the environment and citizen's duty for the protection and preservation of natural flora and fauna respectively. These Constitutional spaces provide for exploring elements that would help in developing a regulatory or institutional framework at the national level on water use efficiency. Needless to say, the Constitutionality of a Central legislation, if at all it needs to be formulated, has to be further explored.

We have also seen that there is no legal instrument at the national level that could be used for introducing systemic efficiency across all sectors. The Water (Prevention and Control of Pollution) Act, 1974, has a specific mandate and there are serious implementation issues with this law, rendering it almost toothless, contrary to the objective and purpose for which it was enacted. The institutional mechanism for the control of pollution under the Water Act has not been decentralised, therefore rendering the Pollution Control Boards ineffective and less efficient in checking violations of the law. Thus, even the law that does not directly provide for efficiency norms for water use but has the potential to be used indirectly also does not stand a chance due to its limited approach in terms of actual implementation.

---

11. See Directive Principles of State Policy, Constitution of India, 1950.

The EPA is an overarching law that has significant provisions for the creation of authorities or institutional structures for protecting the environment but it does not specifically address the efficiency component as one of the aspects of environment protection that needs regulation. In this scenario, it would be prudent to say that the legal or regulatory framework that could support WUE is currently missing.

Lack of Coordination in Different Agencies Involved in Water Management and Governance Leads to Inefficient Institutional Mechanism to Achieve Water Use Efficiency

It is acknowledged in the water policy itself that water, as a resource, is a single, indivisible entity. Rainfall, river waters, surface ponds and lakes and groundwater are all part of one system. Despite this, we practice a fragmented approach to its management under different authorities without there being any coordination or convergence of objectives. Irrigation is entrusted to the Ministry of Water Resources (MoWR); pollution control is handled by the Ministry of Environment and Forests (MoEF); the task of providing safe drinking water is dealt with by the Ministry of Rural Development (MoRD) or the Ministry of Housing and Urban Poverty Alleviation (MoHUPA); small hydro power projects come under the Ministry of New and Renewable Energy (MNRE) and; large dams are with the Ministry of Power.

There is a water quality assessment authority which sets standards for potable drinking, irrigation and industrial usage. This authority should have the power to determine as to which water is to be used for what purpose. Once this authority has declared the water fit for drinking, it must never be diverted to other uses. The Central Water Commission, National Water Academy, Central Ground Water Board and the Central Water and Power Research Station all have to work in tandem. The objectives of the 'National Perspective Plan' (NPP), prepared jointly by MoWR and the Central Water Commission, Command Area Development & Water Management, Bharat Nirman, accelerated rural water supply programme which has been renamed as National Rural Drinking Water Programme (NRDWP), have to be harmonised or at least a modicum of interaction should be ensured to create coordination among these agencies, as they are all engaged in the same task of providing rural water supply. Given the enormity of the water challenge and prospects of increasing water scarcity, there is a need to have convergence of various policies and programmes, preferably under one comprehensive legal instrument that provides for the enhanced coordination between different agencies, institutions and authorities working on different but interdependent and interrelated aspects of water management.[12]

Introducing Efficiency in Water Use would Require a Dedicated Institutional Setup which is Currently Missing

It's a matter of further inquiry as to what shape the regulatory aspects of WUE would take. There are a couple of options that can be explored. Bringing normative standards for water intensive sectors with strong monitoring mechanism entrusted to one of the institutions among the present set of agencies involved in water management is one option. The other option could be creation of a new institutional setup under a separate legislation or regulation using the strength of Article 252 or Section 3 (3) of the EPA. These seem to be necessary and viable options given the current form of complexity in water governance in the country today. Currently, different aspects of water management come under different programmes of the Central and state governments. These range from facilitating drinking water supply in rural areas to developing different participatory models for water governance. Thus, for example, the Accelerated Rural Water Supply Programme and the Rajiv Gandhi National Drinking Water Mission (RGNDWM) are run by the Central government as well as state-level departments for rural development.

Similarly, state-level water resources departments and various other state-level departments such as public works, public health department and irrigation are involved in the implementation and monitoring of various water related programmes. As can be seen very clearly, there is an involvement of various ministries and departments for various water related services and its management. So far, efficient water use is not taken up by any of the involved agencies probably due to the reason that making water available to all is a challenge by itself. The presence of many agencies for handling water for various purposes is a justification of the nature of the problem in the eyes of the Government of India. Therefore, currently we

---

12. Like the one issued for MNREGA and RGNDWM.

are not aware of any institutional innovation or effort for bringing efficiency in water use.

Directions for Water Use Efficiency are Missing under the Policy Instruments

Policy instruments in India have persuasive and visionary value only. They are not justifiable. This is the major reason as to why the most forward looking objectives in water policies of 1987 and 2002 have not seen the light of the day. States have also not come forward to bring policy instruments which would at least get the vision of the government in place for managing its water resources. The National Water Policy of 2002 does not address the issue of efficient utilisation of available surface and groundwater resources. However, the policy is under review for achieving the targets of the Twelfth Five Year Plan. Other sector-specific policies such as Industrial Policy, Special Economic Zone Policy, Forest Policy, and National Hydropower Policy do not address the issue of efficiency in utilising water resources. In the wake of an emerging water crisis at the global and local level, it is high time that all these water intensive sectors become accountable to water use.

Water Pollution Laws Silent on Water Use Efficiency and therefore Industries go Unregulated for Exploiting Water Resources

The Water (Prevention and Control of Pollution) Act was enacted in 1974 to check the menace of pollution and for maintaining or restoring the wholesomeness of water in the country. Similarly, the Water (Prevention and Control of Pollution) Cess Act was enacted in 1977 to make it compulsory for those carrying on certain types of industrial activities to pay a cess on water consumption. The revenue thereof was to be used by the Central Board and the State Boards for its anti-pollution tasks, as provided in the 1974 Act. While these legislations deal with pollution and its various sources, neither of them touch upon the issues of conservation and efficient use by the industry and commercial establsihments. As far as agriculture as one of the most water intensive sectors is concerned, there is a big gap between the irrigation potential and its use, as substantiated in studies by IIMs and sponsored by MoWR. Many of the irrigation projects in the country have also been under operation below their potential due to inadequate maintenance, which is one of the important factors for reduced irrigation efficiency at the project level. This has resulted in the problem of low efficiency of water usage and low productivity.

The National Water Mission under the National Action Plan for Climate Change provides Necessary Mandate for Formulating Water Use Efficiency Regulatory and Institutional Framework

The National Water Mission, under the National Action Plan on Climate Change, envisages an increase in the efficiency of water usage by 20 per cent. The mission envisages having different strategies for consumptive and productive uses of water. The objective can be achieved by ensuring improved efficiency, both on the demand side as well as the supply side. Similarly, full utilisation of the created facilities, better design and proper operation and maintenance would considerably help in improving the efficiency on the supply side. The use of micro irrigation, promotion of water neutral and water positive technologies, recycling of water, etc., would also prove very efficacious. Some of the other strategies outlined by the water mission are labelling of water appliances and fixtures, promoting a mandatory water audit (drinking water included), incentivising recycling of water including wastewater, giving awards for conservation-cum-efficient use, incentivising the use of efficient irrigation practices and the full utilisation of the created facilities.

In the light of the above, the government needs to take appropriate measures for improving water use efficiency in various sectors through action plans supported by statutory norms. Improving water efficiency will not only help India reduce water scarcity and improve the water infrastructure, but will also lead to the maximisation of water infrastructure and reduced environmental degradation. In rural areas, the new efficiency concept must be infused into the common and traditional way of using water resources. It is therefore required that the government explore different ways to enhance water use efficiency and find ways to maximise the value of water use and allocation decisions within and between sectors for the sustainable social and economic development of the country.

References

Ministry of Environment and Forests (2009). *Impact of Climate Change on Water Resources in India. www.envirofor.nic.in*

Singh, Chattrapati (1992). "Water Law in India", *Water Project Series.*

11 | Groundwater Legislation in Pakistan

SHAH BAKHT SOHAIL

The dynamic interaction between groundwater and surface water is governed by the hydrogeological principle of Darcy's Law, which explains that 'when an aquifer is hydraulically connected to a stream, the flow into or out of the stream is proportional to the difference between the stream stage elevation and water table elevation.'[1] Groundwater in Pakistan has become a major source not only for agriculturists, but also for domestic users, the latter estimated to be as much as 70 per cent of the population.[2] Furthermore, it is estimated that since 1976, owing to restraints on large-scale surface water exploitation, the agriculturists took recourse to groundwater through the installation of more than 600,000 private tube wells. It is further estimated that 75 per cent of the increase in water supplies in the last three decades owes itself to public and private groundwater exploitation. In addition, many industries use the relatively clean groundwater to meet their water needs.[3]

What is of concern, however, is that despite the importance of the groundwater resource, it continues to be taken for granted, without any thought being given to its systematic monitoring and management. In two of the four provinces of Pakistan, there are no laws to speak of for regulating the use of groundwater. In Baluchistan and Sindh, even though a regulatory legislation is in place, it continues to await implementation. This paper looks at the existing pattern of groundwater rights allocation in Pakistan and suggests the course of action that should be taken by the authorities responsible for devising the laws for the exploitation and

---

1. Washington Water Law Treatise. p.V:2.
2. Steenbergen and Gohar (n.d.): 2.
3. Ibid.

preservation of groundwater. In doing so, this paper also reflects upon the problems and issues that are part and parcel of the overexploitation of groundwater.

Legislations in Pakistan on Groundwater Exploitation[4]

The first piece of legislation for the management of groundwater was the Punjab Soil Reclamation Act, 1952, which was promulgated to control waterlogging and salinity caused by the excess construction and utilisation of drainage tube wells. A Soil Reclamation Board was formed under this Act, which was not only empowered with groundwater management tasks but also to 'instigate a licensing procedure, permitting land owners to install private tube wells.'[5] Later, when the Board was suspended, its executive powers were transferred to the Provincial Irrigation and Power Department. The department framed the licensing rules first in 1965, but they were never enacted. The 1958 legislation, the Pakistan Water and Power Development Authority Act, was enacted to cover the same grounds, creating the Water and Power Development Authority (WAPDA). According to this Act, WAPDA, a federal agency, had control over the country's underground water resources and could issue area-specific orders for its use. These rules, again, were never enforced. The next effort was made by the Government of Balochistan in 1978, with the enactment of the Groundwater Rights Administration Ordinance, which was introduced to control groundwater mining. Under this ordinance, a procedure was laid down to issue permits for digging new *karezes*, i.e., wells and tube wells, but 'specific guidelines for each region were never issued and [... the] Ordinance did not make a noticeable impact on the groundwater rush that continued unabated' during the next three decades.

The Environment Protection Agency Act, 1996, for the first time voiced concern over the poor quality of surface and groundwater in the country. The federal and provincial Environment Protection Councils are supposed to set the quality standards but, so far, they have not dealt with the groundwater quality.

---

4. Ibid.

5. Ibid.

Lastly, the Provincial Irrigation and Drainage Authorities Act, 1997 (replaced in Sindh by Sindh Water Management Ordinance, 2002), provides the basis of the formation of Provincial Irrigation and Drainage Authorities. These authorities, entrusted with irrigation and drainage management, are supposed to ensure that the groundwater monitoring is undertaken and have been mandated to 'initiate policies to address groundwater management problems.' So far, nothing of the sort has been done. In Punjab and Khyber Pakhtunkhwa, these authorities are largely invisible, not to speak of being functional.

All of Pakistan's legislation related to groundwater management, however insignificant, has a common denominator: it somehow fails to get implemented.

According to Steenbergen and Gohar (n.d.), the groundwater policy of Pakistan can be summarised as follows:

> The focus of groundwater policy was on control of water logging and the stimulation of private tube well development. Government interventions were supply driven. There was little concern about overexploitation of aquifers or deteriorating groundwater quality.
>
> Groundwater policy relied on public investment and subsidies rather than on regulation. Legislation was not enforced. It was not clear whether the federal government or the provincial governments should regulate groundwater exploitation. All policies were initiated or implemented at the federal and provincial level. The involvement of the local government was minor. There was no involvement of local farmers' organisations. Local governments and farmers' organisations are weak in Pakistan.

Problems with Overexploitation of Groundwater

Due to the inadequate provision of drainage facilities during the development of the canal irrigation system in the Indus Basin, waterlogging increased throughout the Basin. Associated with waterlogging was the phenomenon of salinisation of the soil from capillary rise and evaporation of the mineralised groundwater. Waterlogging and salinity left large tracts the farmland unfit for irrigating. The Government responded with a crash programme known as the Salinity Control and Reclamation Projects (SCARP) to reclaim the land, which entailed combating high water tables primarily through vertical drainage. 'In 1954, the large tube well drainage scheme started and soon a second benefit of the drainage tube wells was realised—where the tube wells pumped [up] fresh groundwater, they not only lowered groundwater tables, but also augmented surface irrigation supplies.' In much of the affected areas, the programme did wonders, converting the saline land into fertile land. In fresh groundwater areas, the SCARP drainage tube wells doubled as an additional source of water. Also, the government in several ways supported private groundwater exploitation. Foremost was the provision of heavily subsidised power supply to tube well owners. For tube well owners in Punjab and Sindh, the electricity costs were 40 per cent less than the normal rate and in Balochistan and NWFP it was as much as 60 per cent less. Tube well development was further promoted through a series of government programmes, such as providing pumpsets, wells and tube wells for free or on soft-term loans. However, groundwater is not an inexhaustible source, as some fondly believed initially. It is now depleting fast and preserving it is, indeed, a challenge. The overexploitation of the groundwater during the SCARP programme, as well as afterwards—what with the government giving it the full steam ahead with subsidies and easy loans—has led to the reverse problem of falling water tables in Punjab. Punjab, in this sense, is a paradox: it has places without tube wells where the water table is so high that it leads to waterlogging and, in contrast, there are areas where the groundwater has been so overexploited that it has led to a sharp decline in water table, leaving the area unfit for irrigation.

According to Steenbergen and Gohar (n.d.), '...the spectacular increase in the number of private tube wells has changed the setting entirely and invalidated the old policies. In several fresh groundwater areas in the Indus Plain, there has been a complete volte-face. Where 30 years ago high groundwater tables were the major threat, groundwater levels have now declined due to private tube well development. Salinisation through capillary rise is no longer a threat, but the intense pumping poses other threats to soil and water quality.'

Outside the canal-irrigated areas of the Indus Plain, the promotion of private tube well development is outdated. Groundwater is no longer the seemingly limitless resource it once was. In the absence of effective regulation, the large numbers of tube wells have resulted in groundwater mining. In addition, the social issue of distributing the access to groundwater has come to the fore.

The main challenges in groundwater management in Pakistan are five-fold:

(i) Addressing quality deterioration.

(ii) Protection from pollution.

(iii) Reversing the continuous lowering of groundwater tables, especially in the *barani* areas and the canal commands in Punjab.

(iv) Addressing the waterlogging problem, particularly in Sindh.

(v) Addressing the issue of accessibility for those who do not own tube wells, and

(vi) Resolving the financial and institutional problems.

These problems call for a reduction in groundwater extraction, carefully balancing the combination of pumping, irrigation and drainage, and control of groundwater pollution.

Right now, due to the absence of any accountability mechanisms and effective legislation, landowners are free to drill and pump at their will. There is a set of questions that need to be asked. What is the quantum that can be exploited? What are the specifics of the required law, as also of the institutional framework that is needed and what are the necessary measures needed in specific cases of legal and administrative bottlenecks?

According to J.M. Otto, in *Groundwater Law and Administration in Developing Countries*, the answers that have attracted general consensus are as follows:

1. There must be a legal limit to the freedom of landowners to drill and pump at their will.

2. There must be a lower groundwater institution (LGI) at the regional or local level that would: (a) register all wells and users in its area; (b) regulate and reallocate rights to drill and use groundwater, either through a licensing system or some other method and (c) conduct direct monitoring and supervision.

3. There must be a higher (national or provincial) groundwater institution (HGI) which should be given the task and power to: (a) collect and analyse hydrological data on groundwater; (b) carry out

a classification of areas according to the urgency of groundwater problems and specify the kind of management regime needed; (c) coordinate its policy with other higher government institutions and oversee groundwater management at a higher level.

However, the landowners will resist these restrictions, as they consider it their right to use their land [and the water under it] in whatever way they like. Literature on Common Law suggests that any such restriction would be against the law. None of the four Common Law doctrines support state control over the freedom of landowners to use and sell the groundwater they extract from underneath their own property. The fundamental points of the four doctrines in relation to the subject under discussion are briefly mentioned below.

1. *The riparian doctrine or absolute ownership doctrine:* The oldest of the common law doctrines, it scarcely puts any limits on the right of the landowners to extract groundwater, implying that this right is absolute.

2. *The doctrine of reasonable use:* Riparian doctrine is read in combination with this doctrine. In this, the groundwater rights are limited to 'reasonable use' on overlying land.

3. *The doctrine of correlative rights:* Propounded in the United States, it states that in times of scarcity, the court, if called upon, can permit the overlying owner, as correlative or co-equal owners [with the state], to have access to his proportionate share of the groundwater. Thereby, the doctrine empowers the court to restrict the right of proportionate share with the conditionality of reasonable use.

4. *The doctrine of prior appropriation:* Since the three previous doctrines attribute rights to landowners, this doctrine was propounded by the common law jurists to secure the rights of the tenants and landless labour, especially when they have had access to groundwater in the past. This doctrine states that the first water rights are vested in those who were the first users, regardless of who owns the land at present.

Therefore, limits have already been set according to the provisos of 'reasonable use' and 'proportionate share' clauses in the Common Law.

In the Dutch Civil Code, it is stated that 'the property of land includes, unless the law states otherwise, the groundwater that has come to the surface by a source, well, or pump. The legislator assumes that groundwater is res nullius (nobody's thing) until it has been brought to the surface. In other words, until it is pumped up to the surface, it is no one's property. Whether a landowner is allowed to extract it, and how, and in what quantity, can certainly be decided by administrative law. Moreover, the groundwater will become the landowner's property "unless the law states otherwise."' Therefore, in this continental code, there is no legal problem in groundwater control. Similarly, under Common Law, as discussed above, groundwater becomes the landowner's property through accession, but a limit can be put on its use through a specific rule of the environmental law.

Going by Otto's recommendations stated above, the Pakistan Government needs to formulate a clear comprehensive and effective legislation on the issue in hand. One way to do so would be to revive the Groundwater Rights Administration Ordinance, 1958, in Balochistan and to introduce similar legislations in the other provinces. Under this legislation, there should be a rule to enforce licensing for digging new tube wells or extracting groundwater in any part of the country. The basis for issuing these licences/permits should be laid down very clearly, so that there is little room for confusion or for challenging the writ. Also, it must be kept in mind that the effective implementation of any such legislation very much depends on the support and willing involvement of those who would be directly affected by the law, that is, the people—the users.

One legislation of great merit is that of the state of Washington in USA.[6] In 1945, the state extended the 1917 Surface Water Code with a Ground Water Code, thereby bringing it under the ambit of the permit system. By this enactment, the state legislature had thought it fit to reject the 'correlative rights' and the 'reasonable use' doctrines so as to extend the prior appropriation and beneficial use principles of the surface water code to groundwaters. By this move, the legislature also openly revealed its belief

---

6. Christine *et al.* (2000).

in the public ownership to such waters, as against private ownership. In defining the management of this public resource, the legislature made the acquisition rights dependent on compliance with the exclusive permit system of all waters, surface or ground. It should be noted, however, that the enumerated use of relatively minor amounts are exempted from the permit system. The 1945 code was a comprehensive one that did not do away with the vested rights that existed prior to the enforcement of the code. For instance, the code provided for the issuance of water rights certificates to 'any person, firm, or corporation claiming a vested right to withdraw public groundwaters of the state by virtue of prior beneficial use of such water.' To obtain a certificate, however, the code laid down a comprehensive procedure that had to be complied with within three years of the enactment of the Code. The Code, in unambiguous terms, made a permit mandatory post-1945 enactment of the law. The code is comprehensive enough to provide four major classes of exemptions under the groundwater right permit system:

1. Stock watering purposes.

2. The watering of a lawn or of a non-commercial garden, whose area did not exceed the fixed limit.

3. Single or group domestic use not exceeding 5,000 gallons a day, and

4. Industrial use not exceeding 5,000 gallons a day.

According to the code, the users who qualified under one or more of the exemption rules to withdraw water regularly (and were using it beneficially) were entitled to permits with rights equal to those issued under the provisions of the code. Due consideration was also given to instances where small withdrawals might affect the water system.

Since groundwater is now recognised as a depleting, limited resource and utmost care must be taken to keep a check on its use, there is a pressing need to put necessary legislations in place and implement it effectively. The legislature should make sure that laws enacted invite the participation of those affected, starting from the lowest level of the social hierarchy, as it is the only way to make the laws effective. Without this all inclusive involvement, the legislation would be an exercise in futility, as were those that preceded it. What is needed in all the provinces of Pakistan are legislations on the lines of the 1945 Washington Code, which must also enforce the participation of the local governments.

References

Bhatia, Bela (1992). "Lush Fields and Parched Throats: Political Economy of Groundwater in Gujarat", *Economic and Political Weekly* 27(51/52): A142-A170 (December 19-26). Published by: Economic and Political Weekly. Article Stable URL: *http://www.jstor.org/stable/4399241*

Bibliography of Asian Studies (1967). "Pakistan", *The Journal of Asian Studies* 27(5): 252-260 (September 1968). Published by: Association for Asian Studies . Article Stable URL: http://www.jstor.org/stable/2942831

Christine O. Gregoire, James K. Pharris, P. Thomas McDonald (2000). *An Introduction to Washington Water Law.* Published by: Office of Attorney General. Article Stable URL: *http://www.atg.wa.gov/uploadedFiles/Home/About_the_Office/Divisions/Ecology/Intro%20WA%20Water%20Law.pdf*

Dhawan, B.D. (1991). "Developing Groundwater Resources: Merits and Demerits", Economic and Political Weekly 26(8): 425-29 (February 23). Published by: Economic and Political Weekly. Article Stable URL: *http://www.jstor.org/stable/4397370*

Easter, K. William (2000). "Asia's Irrigation Management in Transition: A Paradigm Shift Faces High Transaction Costs", *Review of Agricultural Economics* 22(2): 370-388 (Autumn-Winter). Published by: Oxford University Press on behalf of Agricultural & Applied Economics Association. Article Stable URL: *http://www.jstor.org/stable/1349800*

Egboka, B.C.E., G.I. Nwankwor, I.P. Orajaka and A.O. Ejiofor (1989). "Principles and Problems of Environmental Pollution of Groundwater Resources with Case Examples from Developing Countries", *Environmental Health Perspectives* 83: 39-68 (November). Published by: Brogan & Partners. Article Stable URL: *http://www.jstor.org/stable/3430648*

Feinerman, Eli and Keith C. Knapp (1983). "Benefits from Groundwater Management: Magnitude, Sensitivity, and Distribution", *American Journal of Agricultural Economics* 65(4): 703-710 (November). Published by: Oxford University Press on behalf of the Agricultural & Applied Economics Association Article Stable URL: *http://www.jstor.org/stable/1240458*

Hodgson, Stephen (n.d.). *Modern Water Rights, Theory and Practice.* Published by: FAO Legal Office. Article Stable URL: *ftp://ftp.fao.org/docrep/fao/010/a0864e/a0864e00.pdf*

McFarland, James W. (1975). "Groundwater Management and Salinity Control: A Case Study in Northwest Mexico", *American Journal of Agricultural Economics* 57(3): 456-62 (August). Published by: Oxford University Press on behalf of the Agricultural & Applied Economics Association. Article Stable URL: *http://www.jstor.org/stable/1238408*

Otto, J.M. (n.d.). "Groundwater and Administration in Developing Countries". Supporting Papers. Article Stable URL: *citeseerx.ist.psu.edu/viewdoc/download?doi=10.1.1.131.9572*

Steenbergen, Frank van and M. Shamshad Gohar (n.d.). Groundwater Development and Management in Pakistan: Experiences from Developing Countries.

Steenbergen, Frank van and W. Oliemans (n.d.). *Groundwater Resource Management in Pakistan: Experiences from Developing Countries.* Article Stable URL: *http://www2.alterra.wur.nl/Internet/webdocs/ilri-publicaties/special_reports/Srep9/Srep9-h6.pdf*

Section IV

Sharing Water Resources

12

Safeguarding South Asia's Water Security

The Twin Imperatives of Effective Transnational Water Arrangements and Internal Water Management

MICHAEL KUGELMAN

In today's era of globalisation, the line between critic and hypocrite is increasingly getting blurred. Single out a problem in a region or country other than one's own, and one risks triggering an immediate—yet understandable—response: Why criticise our problem, when you face the same one back home?

Such a response is particularly justified in the context of water insecurity, a dilemma that afflicts scores of countries—including this author's United States. In the parched American West, New Mexico has only 10 years' worth of drinking water left, while Arizona already imports every drop it consumes. Less arid areas of the country are getting increasingly water-stressed as well. Rivers in South Carolina and Massachusetts, lakes in Florida and Georgia, and even the mighty Lake Superior (the world's largest freshwater lake) are all running dry. According to the US Environmental Protection Agency, if American water consumption habits continue to go unchecked, 36 states will have water shortages within the next few years.

Also notable is the fact that America's waterways are choked with pollution, and that nearly 25 million Americans may fall ill each year from contaminated water. Not to mention that more than 30 American states are fighting with their neighbours over water.[1]

---

1. See Barlow (2008); Duhigg (2009) and Glennon (2009). For one of the best contemporary accounts of the US water crisis, see Glennon (2009a).

South Asia's Perilous Water Security

Such a narrative is a familiar one, because it also applies to South Asia. However, in South Asia, the narrative is considerably more urgent. The region houses a quarter of the world's population, yet contains less than 5 per cent of its annual renewable water resources. With the exception of Bhutan and Nepal, South Asia's per capita water availability falls below the world average.[2] Annual water availability has plummeted by nearly 70 per cent since 1950, and from around 21,000 cubic metres in the 1960s to approximately 8,000 in 2005. If such patterns continue, the region could face 'widespread water scarcity' (that is, per capita water availability under 1,000 cubic metres) by 2025.[3] Furthermore, the United Nations, based on a variety of inputs—including ecological insecurity, water management problems and resource stress—characterises two key water basins of South Asia (the Helmand and Indus) as 'highly vulnerable'.[4]

These findings are not surprising, given that there are many drivers of water insecurity in South Asia: High population growth, vulnerability to climate change, arid weather, agriculture-dependent economies and political tensions. This is not to say that South Asia is devoid of water security stabilisers; indeed, its various transnational arrangements, to differing degrees, help the region manage its water constraints and tensions. This paper argues that such arrangements are vital, yet also incapable of safeguarding regional water security on their own. It asserts that more attention to the demand side of water management within individual countries is as crucial for South Asian water security as are transnational water mechanisms.

The Imperative of Transnational Water Arrangements: Water Connectivity in a Poorly Integrated Region

To understand the importance of transboundary water arrangements in South Asia, one must first bear in mind a paradox: the region is poorly integrated, yet linked together by water co-dependencies.

---

2. United Nations Environment Programme (UNEP) and Asian Institute of Technology (2009).

3. Jaitly (2009).

4. UNEP and Asian Institute of Technology (2009).

Consider, first, that the World Bank has declared South Asia the world's least integrated region. According to the Bank, South Asia has the world's worst railways and road density and only the sub-Saharan African region has worse electricity and sanitation systems. Predictably, intra-regional trade accounts for just five per cent of the region's total international trade, and less than two per cent of gross domestic product.[5] Not surprisingly, South Asia's regional organisation, SAARC, is not nearly as dynamic as regional groupings like the European Union or the Association of Southeast Asian Nations (ASEAN).

Consider, second, that the region's countries depend on the same rivers—and, by extension, neighbouring upper riparians—for their water supply. India, Bangladesh and Nepal look to the Brahmaputra, while both Pakistan and India are beholden to the rivers of the Indus Basin. Bangladesh and Pakistan, both lower riparians, must obtain great majorities of their water resources (91 and 75 per cent, respectively) from beyond their borders.[6] Conversely, China, while not a geographic entity of South Asia, is an upper riparian for many of the key rivers flowing into South Asia. India is both a lower (in the case of the Brahmaputra) and upper (in the case of the Indus) riparian. This means, hypothetically, that any flow-diverting Chinese activities on the Brahmaputra would alarm India, Nepal and Bangladesh, and in turn trigger Indian flow manipulations on the Indus, with implications downstream for Pakistan. Such a dynamic creates a hydro domino effect: one nation's water policies can spark a chain reaction throughout the region.

Another notable factor about South Asia's interconnected water geography is that many major rivers originate in or pass through politically contested or tense areas. The Tibetan Plateau—where the mighty flows of the Salween, Brahmaputra, Indus, Sutlej and other rivers all spring to life, providing water to 1.5 billion people downstream—is controlled by China, and abuts India's water-rich Arunachal Pradesh state, which China covets and has sparked Sino-Indian tensions. The rivers of the Indus Basin, of course, flow through the Kashmir region—an unending source of Pakistan-India tensions. No wonder that many of South Asia's riparian pairings

---

5. World Bank (2006) and Kugelman (2008).

6. Asia Society (2009).

(India-Pakistan for the Indus, and India-Bangladesh for the Brahmaputra and the Ganges) reflect the region's most troubled bilateral relationships.

In effect, South Asia's water relations play out amid a volatile backdrop of shortage, dependency and geopolitical tension.

Reactions with Regional Ripple Effects: Upper Riparians

A nation's upper riparian status by no means guarantees water security. China, an uber-upper riparian, suffers from a full-blown water crisis. The North China Plain—one of the country's 'economic and social cores,' generating more than 20 per cent of its grain supply, according to China scholar David Pietz—is frighteningly water-scarce, with a per capita availability of 225 cubic metres per year.[7] India, meanwhile, contains about 20 per cent of the world's population, but only about 4 per cent of its water.[8] A 2010 Asian Development Bank report projects that the country could suffer from water shortages of 50 per cent by 2030.

Faced with current shortages, policymakers in upper riparian states frequently opt for supply-side solutions. Water-generation measures may ease water stress internally, yet they often exacerbate regional tensions. India and China, to generate desperately needed water resources for both agriculture and energy, often resort to dam construction and other large engineering projects. Many of these are run-of-the-river and hence do not prevent flows from continuing downstream. Still, dams are a delicate matter. China's insistence that its South to North Water Diversion Project will not divert flows from the Brahmaputra is met with scepticism in India. Additionally, even run-of-the-river dams and other hydro projects threaten lower riparians' water and food security. For example, if India builds all its envisioned hydro projects on the western rivers of the Indus Basin, Pakistan's agriculture could be deprived of up to a month's worth of river flows—enough to ruin an entire planting season.[9]

---

7. Pietz (2010).

8. Specter (2006).

9. Polgreen and Tavernise (2010).

Reactions with Regional Ripple Effects: Lower Riparians

Lower riparians also opt for large, supply-generating projects. Pakistan's water resources policy has been dominated by dam-building for decades, and the country's Water and Power Development Authority (WAPDA) plans to construct five dams, three large canals and five hydropower facilities by 2025. As lower riparians, these nations' engineering projects will not impede river flows into downstream countries. However, they upset riparian communities, who risk being displaced. Supply-side efforts also aggravate regional and provincial tensions, which are high in many South Asian countries. The furious debate surrounding Pakistan's Kalabagh Dam, for example, has pitted supporters in Punjab against opponents everywhere else in the country. This internal discontent feeds into the domestic instability that so concerns Pakistan's neighbours.

In Pakistan, water insecurity also spawns different manifestations of militancy. In recent years, the Pakistani Taliban, aware of Pakistan's precarious water situation, has attacked the country's largest earth-filled dam, the fabled Tarbela. More recently, extremists in [Pakistan's] Punjab have issued violent threats, angrily blaming India for 'stealing' Pakistan's water and vowing aggression against India unless it ceases such 'theft'. It is perhaps not a coincidence that while Pakistan-based extremists have been largely absent from the latest uprising in Jammu and Kashmir, they have been increasingly vocal in their accusations of India's alleged water theft. These extremists regard their India blame game as an expression of nationalism—an attempt to draw attention away from divisive debates within Pakistan about dam construction, and toward the need for putting up a unified front to confront India.[10] Water may well be supplanting Kashmir as these militant chiefs' rallying cry.

Another response—more potential than actual, at least at this point—is for desperate citizens to flee to more water-secure nations. According to a Strategic Foresight Group estimate, water scarcity will contribute to the displacement and migration of 50 to 70 million people in India, Bangladesh, Nepal and China by 2050.[11] To be sure, water strain takes a

---

10. Smith (2010).

11. Strategic Foresight Group (2010).

devastating toll on human security inside upper and lower riparian nations alike—as evidenced by the livelihood-shattered Pakistani fishermen along the disappearing Indus Plain and the water-starved, suicidal Indian farmers in Chhattisgarh.

However, migration is a particularly worrisome threat for lower riparians, with the potential not just to uproot millions of people, but also to imperil regional stability. Bangladesh offers a vivid example. As an impoverished, highly populous lower riparian that depends on other countries for nearly all its water needs, it is deeply vulnerable to Indian or Chinese activities on rivers upstream and to glacial melting in the Himalayas. And as a low-lying nation, it is susceptible to rising sea levels and monsoon flooding. This array of water problems, according to some observers, could hasten mass Bangladeshi migrations into India's volatile east, with politically explosive implications.[12]

Nightmare Scenarios

The combustible mix of water vulnerability, geopolitical tensions and supply-side responses with ripple effects across borders constitutes a recipe for disaster. In the coming years, these ingredients could simmer to a boil as the region's population growth soars and Himalayan glacial melt accelerates. Several nightmare water-driven scenarios come to mind:

- *Indus Basin war*: Militants in eastern Pakistan, vowing to avenge India's 'theft' of Pakistan's water from the Indus Basin's western rivers, launch terror attacks in India. New Delhi dispatches troops to its western flank, and threatens to shut off western river flows into Pakistan. With India-Pakistan relations having improved in 2011, this scenario may well be averted—yet by no means can it be ruled out.

- *Sino-Indian water showdown*: China, in response to Indian defensive upgrades in and near Arunachal Pradesh, seeks to reclaim a strategic advantage by slowing the flow of the Brahmaputra into India's Assam state—an impoverished bastion of

---

12. Friedman (2009).

separatist militancy that is also an important area of agricultural production.

- *Environmental refugee crisis*: As glacial melt in the Himalayas runs its course, river flows slow to a trickle, and lower riparian Bangladesh experiences rampant water scarcity. Bangladeshis migrate en masse to more water-secure but politically volatile eastern India—deepening instability in the latter as violent factions target these new arrivals, and long-entrenched separatist militants exploit the unrest by launching their own attacks.

These hypothetical scenarios, and the bubbling cauldron of water insecurity and political tensions that makes them impossible to dismiss, crystallise the importance of transboundary water arrangements with the ability to manage, if not reduce, the region's water-based tensions.

South Asia's Transnational Water Arrangements: Victories and Vulnerabilities

Numerous transboundary rivers in South Asia are governed by treaties. These include the Mahakali (to which India and Nepal are party), the Ganges (involving India and Bangladesh) and the Indus (comprising India and Pakistan). The Mahakali accord is meant to promote the river's integrated development, though various disagreements have prevented the treaty from being properly implemented. The Ganges agreement is also constrained by several disagreements, particularly over the Farakka Barrage, which Bangladesh believes has reduced downstream flows of the Ganges.

The Indus accord, however, has had considerable more success.

Indus Waters Treaty: Success Story

The Indus Waters Treaty (IWT) allocates the Indus Basin's western rivers to Pakistan, and the eastern rivers to India (India is authorised to draw on the western rivers for agricultural purposes, so long as this use does not involve storage). It deserves particular attention, given that it is repeatedly lauded as a magnificent example of international cooperation and conflict management. Such rhetoric is warranted. The two parties, despite their rocky relations, truly make a commitment to uphold the treaty's provisions. The IWT emphasises transparency—particularly in terms of data exchange and notification of plans to undertake hydro projects. Sure enough, in 2010, India allowed Pakistan to inspect several under-construction Indian hydropower projects on the western rivers. The two nations have also agreed to set up a telemetry system to measure river flows.

Additionally, only once in the treaty's 50-year history have the accord's mechanisms for dispute resolution been put to the test. This was a period between 2005 and 2007, when a neutral expert (appointed by the World Bank, as per the IWT's stipulations) weighed in on the concerns of Pakistan about the Baglihar Dam, a project being constructed by India on one of the Indus Basin's western rivers. The expert sided against Pakistan on several technical points—particularly when he ruled that a gated spillway was necessary—yet agreed with it on others (such as on the issue of power intake levels). While many on both sides undoubtedly find ways to oppose the outcome, it is undeniable that the dispute was settled peacefully. In June 2010, both governments concurred that the Baglihar dispute had been definitively resolved.

In 2010, Pakistan decided to bring some technical concerns about the Kishanganga Dam (a hydro project being constructed by India in Jammu and Kashmir) to an international court of arbitration. While the outcome of this latest case is far from clear, the Baglihar precedent suggests the arbitration process for Kishanganga should be similarly smooth.

Indus Waters Treaty: Cracks in the Armour?

Despite its successes, many observers fear the IWT's increasing vulnerability. This derives in part from the long-standing grievances harboured by both countries about the treaty. Indians believe it curtails the storage rights of Jammu and Kashmir, and hampers the development of hydro projects on the western rivers. Indeed, India's participation in the treaty is both economically and politically risky given that the energy-starved nation must forego opportunities to harvest hydroelectric resources, and given the resentment these cautious hydro development policies breed among Kashmiris.[13]

---

13. Smith (2010a).

Pakistan, meanwhile, resents its inherent vulnerability as the lower riparian (even though the IWT allots it 80 per cent of the total Indus Basin river flow). For many in Pakistan, the country with the lowest per capita water availability in Asia, entrusting its water security to its deeply mistrusted neighbour is profoundly unsettling.

Furthermore, worries about population growth and climate change effects (particularly the melting of Himalayan glaciers) are fueling calls for a revised treaty that takes such trends into account. There are also recommendations that the treaty be expanded, so that it includes the other two Indus Basin riparians, China and Afghanistan. Those advocating this latter view argue that a four-party treaty would reduce the possibilities of region-wide conflict, and promote Basin-wide ecological sustainability.[14]

Toward a Resource-Sharing Paradigm?

Some experts contend that the 'dividing the resources' mentality of the IWT is no longer tenable, and that Pakistan and India should move toward a more cooperative 'share the resources' paradigm. Ahmad Rafay Alam, one of the most eloquent articulators of this view from the Pakistan side, argues that the two nations should determine 'whether it is in the economic, social or political interest of both riparians to cooperate on water, rather than be antagonistic over it.'[15] For example, instead of railing against Indian run-of-the-river dam projects on the western rivers, power-starved Pakistan should consider buying the electricity they generate—a more affordable investment than buying power from pricier indigeneous gas-powered rental power projects.

This paradigm can be applied to a broader geographic area as well. Bhutan and Nepal are blessed with ample hydroelectric potential; India could therefore invest in energy resources in these countries just as Pakistan could in India.

Transnational Transparency and the International Community's Role

Some experts make more modest suggestions about improving transnational water arrangements. They believe that regional water security is

14. Strahorn (2010).
15. Alam (2010).

best attained not by changing the architecture or philosophy of regional water agreements, but instead by strengthening mechanisms for cooperation and transparency within existing arrangements. In other words, instead of crafting a new IWT, the parties should make even greater commitments to respect the treaty's current provisions on transparency.

A core component of this viewpoint is the need for more data-sharing on river flows, hydro project plans and glacial melting. This emphasis on transparency is one of the chief features of US government policy on transnational water arrangements. Policymakers in Washington, when describing South Asian water arrangements, repeatedly underscore the same terms: good communication, data-sharing and joint management.

Some modest steps have already been taken along these lines. As mentioned above, India and Pakistan recently agreed to several information-sharing measures. Additionally, China and India have agreed to share data on glacial melt. Still, much more can be done—and it is here that the international community can play a key role. International academic conferences and other forums facilitating the exchange of information about water constitute one way to boost water transparency. Technology-sharing is another avenue. The United States, for example, can share its new metric system—a water measurement tool developed by academic researchers that has already helped resolve a fight over water in the Arkansas River, and could similarly be deployed in the Indus Basin. Similarly, America could provide satellite data that documents water availability. It is sometimes observed that there are no international water treaties or other global mechanisms to promote water cooperation. The international community's participation in water data-sharing exercises in South Asia and elsewhere can serve as a modest corrective.

Champions of transboundary water transparency argue that by fostering greater data-sharing, riparians' mutual suspicions can be reduced, paving the way for greater water cooperation. It is important to note, however, that greater water transparency can easily backfire, and even undermine riparian relations. As one study concludes, revelations of 'inequitable Indian water stewardship on Indus tributaries' could exacerbate the volatility of India-Pakistan relations.[16] One might then argue

---

16. Dabelko and Sticklor (2010).

that nations with delicate relations may in certain cases be better served by not revealing inflammatory water data, and should instead maintain opacity.

This raises a critical point: advocating for more transboundary water openness, pushing for 'resource-sharing' paradigms and proposing more inclusive treaties—all presuppose a level of political cooperation that may not exist in South Asia. After all, it is easy to propose the formation of a new Himalayan Rivers Commission to govern water management in this sub-region, as was done in 2010. However, it would be much more difficult to implement such an arrangement, which would necessitate the close cooperation of nations harboring major trust deficits toward each other.

So the question invariably arises: how can South Asia's transnational water arrangements be enhanced in such a troubled political environment?

One answer is to look within, and to muster better efforts to improve internal water governance. This is because sounder water management inside countries can create a more favourable political climate for the pursuit and achievement of lasting external water cooperation. Better internal water management can serve as a catalyst for effective regional water governance.

The Imperative of Better Internal Water Management

It is common to attribute water problems exclusively to politics. It is often said that India-Pakistan water tensions are just one facet of a long-troubled bilateral relationship. Similarly, one frequently hears that India-China disagreements over water are rooted in the larger competition between the two rising powers for influence over South Asian territory and resources. And India's water disagreements with Bangladesh and Nepal are said to be part of a long history of poor political relations.

Such views are accurate, yet incomplete. After all, across the United States and Australia, regions and states bicker over river water allocations— yet these tensions have little to do with politics. Neither Colorado and Kansas in the United States, nor Victoria and Queensland in Australia, are at risk of going to war; they simply disagree about how to properly divide up river flows. As such, their squabbles demonstrate an inability to efficiently manage existing water resources. In South Asia, where the availability of water resources is more precarious, this poor water governance is the main cause of water insecurity.

Abysmal Water Governance

In South Asia, water insecurity is not solely a function of resource shortages. To be sure, much of it is running low on water. However, excluding some arid portions of the region, very little of South Asia is actually water-scarce. The resource is precious, yet present. South Asia's water problems are very much rooted in the wasteful and inefficient management of the region's available water supplies.

Case Study of Pakistan

Pakistan is arguably the worst culprit. Water infrastructure and transmission systems—the canals and pipes that have helped make the Indus River system the world's largest contiguous irrigated area—are literally falling apart because they have not been properly maintained. As a result, millions of gallons of water are lost to leakage every day. In urban areas, wastewater treatment facilities are well nigh non-existent. Hence, the country's great cities are notorious for fetid surface water resources that sicken and kill hundreds of thousands every year.

Meanwhile, Islamabad offers few incentives to the population to use water-saving technology. The lack of subsidies for drip irrigation, for example, compels farmers to use traditional, water-wasting flood irrigation. Furthermore, the government has made little effort to diversify crop production. This is unfortunate, given that Pakistan's most intensively produced crops—and those that fetch the greatest profits for small farmers—are the most water-guzzling.

Then there are the structural factors that exacerbate Pakistan's water mismanagement. Thanks to the nation's feudal land setup, a small landed elite owns most of the rural land, while the majority of the rural population is landless. In a country with few water laws or rights, land ownership determines water access. As a result, most of Pakistan's rural population struggles to obtain water. In theory, mechanisms such as *warabandi*—the colonial-era water distribution system meant to ensure farmers equal access to irrigation water—and provincial-level water arbiters such as the Indus River System Authority (IRSA) are meant to compensate for such inequalities. In reality, they fail miserably. *Warabandi* is exploited by politically connected large farmers, while IRSA is rarely taken seriously, and its edicts about provincial river flow allocations are routinely ignored.

Given this gloomy domestic water situation in Pakistan, it is clear why not only militants, but also the country's government, have chosen to externalise blame over the border. After all, bringing attention to its dysfunctional domestic water management would essentially be an acknowledgment that Islamabad is to blame for its water crisis. Granted, Islamabad is much less prone to blame India than it was years ago (in fact, in 2010, Pakistan's Minister for Water and Power acknowledged that India rarely prevents river water from flowing into Pakistan), yet it rarely admits that the country's water problems are largely internally rooted.

Case Study of India

Pakistan is not the only poor water manager in South Asia; India's water governance is similarly troubled. India is home to 'dilapidated' pipes and pumping stations that cause more than one-third of New Delhi's fresh water (and at least 40 per cent of most Indian cities' total water resources) to be lost to leakage. The Yamuna River is choked with 'fecal bacteria', and this sewage has increased 'thousands of times' over the last decade. Water-intensive rice and wheat crops are championed by the government through price guarantees to farmers. And per capita storage availability—the *sine qua non* for dam efficiency—has plummeted in recent years to levels found in Africa's poorest nations.[17]

Implications of Poor Internal Water Management

Poor internal water management has grave implications for public health, food security and the environment. Sometimes the effects can be catastrophic. Consider the flooding in Pakistan in 2010. If the country's water infrastructure had been sturdier and better maintained, raging rivers would have been more contained and the damage wrought by the deluge may not have been as extensive.

---

17. Specter (2006).

Strain on Groundwater Resources

Perhaps the most disturbing implication is the strain on groundwater resources. With poor internal water management regimes causing surface water to be wasted, lost, or contaminated across the region, South Asians are increasingly digging deeper—literally—to alleviate their water insecurity.

In the context of agriculture—by far, the sector that consumes the most water across South Asia—the increasing inefficiency of highly subsidised, state-run irrigation systems has driven farmers to mine groundwater, over which they have more control and is more readily available. Even back in 2000, nearly 70 per cent of Bangladesh's irrigation and more than 50 per cent of India's was served by groundwater resources.[18]

Groundwater depletion goes beyond the agricultural sector. According to the World Bank, India is the world's most voracious consumer of groundwater. This heavy consumption is reflected in a 2009 study by several US scientists, which found that groundwater levels fell by about four centimetres per year between 2002 and 2008 across three states in northwestern India—including the breadbasket state of Punjab.[19] These areas could, conceivably, exhaust their entire groundwater supply within the next few decades.

Pakistan, too, is increasingly becoming groundwater-reliant. Lahore—the country's second largest city—is completely dependent on it for drinking water needs, and groundwater tables have fallen by as much as 65 feet in some areas. Worse, wastewater is now infiltrating the city's groundwater, choking it with arsenic.[20]

Once a pristine, untapped resource, groundwater is now being extracted intensively throughout South Asia. In effect, with this onslaught, the last bastion of South Asia's water security has been breached. And as groundwater becomes increasingly short and scarce, the region may be compelled to return to rapidly dwindling surface water resources and to

18. Jaitly (2009).

19. Times of India (2009).

20. Chaudhry and Chaudhry (2009).

compete ferociously for the ultra-precious supply that remains—a terrifying prospect, and particularly so for lower riparians.

Imperilled and Insufficient: Transnational Water Cooperation Amid Poor Internal Water Management

Faced with domestic water problems, and mistaking mismanagement for shortages (or intentionally cloaking mismanagement in the guise of shortages), governments succumb to their supply-side fancies and construct more dams and reservoirs. Such actions, as noted earlier, fan provincial tensions and, in the case of upper riparians, anger downstream neighbours. Herein lies the troubling link between poor water management at home and transnational water cooperation: the former prompts governments to take actions that threaten the latter.

Whether such actions outweigh the risks of imperilling transnational water cooperation is debatable. This is because many supply-side coping strategies are neither efficient nor sustainable. One of Pakistan's top water experts, Simi Kamal, has calculated that the quantity of water projected to be generated by the nation's under-construction Diamer-Basha Dam pales in comparison to the amount that would be freed up simply by repairing and maintaining Pakistan's leaky canal system.[21] Additionally, Pakistan's dams, like India's, are rapidly losing storage capacity.

Such considerations give way to another unsettling reality: so long as internal water management remains poor, the benefits accruing from closer regional water cooperation will be strictly political; from a water-resources standpoint, little will improve. Take the case of Pakistan. Assume, for a moment, that increased cooperation enables Pakistan to succeed in getting India to release more water downstream. What would be the result? Many Pakistanis would argue that their water problems would be solved: parched farmland would be saved, children's thirst would be quenched, and lost water livelihoods would be restored.

In reality, none of this would happen. Instead, more water would mean more inefficiency. More water lost to leaky canals and pipes, wasted in irrigation, showered on water-guzzling crops, and contaminated by urban waste. Indeed, if nothing is done to improve internal water governance,

---

21. Kamal (2009).

allowing more water to gush into Pakistan would simply intensify the country's water crisis.[22]

Recommendations and Conclusion

South Asian nations need to focus more on demand-side solutions to domestic water problems. These include water-conserving technologies, crop diversification, better investments in infrastructure maintenance and wastewater treatment, and a stronger embrace of rainwater harvesting (a conservation method that has already caught on quite strongly in parts of the region). Such policies are less expensive, and potentially more efficient, than traditional supply-side water engineering projects like large dams. Some encouraging signs are emerging from India, where there has been some debate about the merits of emphasising sugarbean cultivation over that of sugarcane, which is notoriously water-wasting. There has also been discussion about embracing water-saving mechanisms such as the direct seeding of rice.

If such demand-side management policies are implemented successfully, South Asian nations would become more judicious in their use of existing water resources, and therefore less threatened in the short term by the spectre of scarcity. Upper riparians would, presumably, be less likely to initiate new hydro-generation projects that upset their downstream neighbours. Lower riparians, meanwhile, would have less incentive (and fewer grounds) to stoke tensions with their upstream neighbours by accusing them of water theft. As a result, transnational water arrangements would be threatened less, and the calmer political climate would enable riparians to make more substantive progress on the data-sharing and transparency essential for better South Asian water security. None of this, it should be noted, would necessitate drawing up new treaties or other water agreements.

To be sure, new demographic and environmental realities may well call into question the continued relevance of decades-old transnational water arrangements. Still, these mechanisms need not stop functioning simply because of the presence of factors that were not at play 50 years ago. One study of the Baglihar Dam case observes that the issue was 'addressed

---

22. Kugelman (2010).

bearing in mind the technical standards for hydropower plants as they have developed in the first decade of the 21st century, and not as perceived and thought of in the 1950s when the [IWT] was negotiated.'[23] A precedent has effectively been set for new conditions to be taken into account when interpreting the existing treaty, without needing to incorporate such conditions into an altogether new or revised treaty.

This is just one more reason for South Asian nations to redouble their efforts to ameliorate internal water management. Transnational water arrangements can also stand to improve, yet they are not in desperate need of reform and revision. Rather, it is the water governance of the region's individual countries that so urgently needs to be fixed. In effect, South Asian water policies must adopt a new approach—one that, in the words of noted water expert Ramaswamy R. Iyer, embraces the 'responsible, harmonious, just, and wise use of water.'[24] With population growth and climate change continuing apace, the stakes have never been higher, and the costs of inaction never starker.

References

Alam, Ahmad Rafay (2010). "New Approach to the Indus Treaty", *The News*, July 23. *http://www.thenews.com.pk/daily_detail.asp?id=252248*

Asia Society (2009). *Asia's Next Challenge: Securing the Region's Water Future-A Report by the Leadership Group on Water Security in Asia.* April. *http://asiasociety.org/files/pdf/WaterSecurityReport.pdf*

Barlow, Maude (2008). "Where Has All the Water Gone?", *American Prospect*, June 12. *http://www.prospect.org/cs/articles?article=where_has_all_the_water_gone*

Chaudhry, Anita and Rabia M. Chaudhry (2009). "Securing Sustainable Access to Safe Drinking Water in Lahore", in Michael Kugelman and Robert M. Hathaway (eds.), *Running on Empty: Pakistan's Water Crisis.* Washington, DC: Woodrow Wilson Centre. *http://www.wilsoncenter.org/topics/pubs/ASIA_090422_Running%20on%20Empty_web.pdf*

Dabelko, Geoff and Russell Sticklor (2010). *Understanding the Security Implications of Water Challenges: India.* Environmental Change and Security Program, Woodrow Wilson Centre. Unpublished Commentary.

Duhigg, Charles (2009). "Millions in US Drink Dirty Water, Records Show", *New York Times*, December 7. *http://www.nytimes.com/2009/12/08/business/energy-environment/08water.html?pagewanted=2&_r=4*

Friedman, Lisa (2009). "Bangladesh: Where the Climate Exodus Begins-Facing the Specter of the Globe's Biggest and Harshest Mass Journeys", *Environment and Energy*

---

23. Salman (2008).
24. Iyer (2010).

Publishing/Climate Wire, March 2. *http://www.eenews.net/special_reports/bangladesh/part_one*

Glennon, Robert (2009). "Our Water Supply, Down the Drain", *Washington Post*, August 23. *http://www.washingtonpost.com/wp-dyn/content/article/2009/08/21/AR2009082101773.html*

———. (2009a). *Unquenchable: America's Water Crisis and What to Do About It.* Washington, DC: Island Press.

Iyer, Ramaswamy R. (2010). "Approach to a New National Water Policy", *The Hindu*, October, 29. *http://www.hindu.com/2010/10/29/stories/2010102963801400.htm*

Jaitly, Ashok (2009). "South Asian Perspectives on Climate Change and Water Policy", in David Michel and Amit Pandya (eds.), *Troubled Waters: Climate Change, Hydropolitics, and Transboundary Resources.* Washington, DC: Stimson Centre. p.22.

Kamal, Simi (2009). "Pakistan's Water Challenges: Entitlement, Access, Efficiency, and Equity", in Michael Kugelman and Robert M. Hathaway (eds.), *Running on Empty: Pakistan's Water Crisis.* Washington, DC: Woodrow Wilson Centre. *http://www.wilsoncenter.org/topics/pubs/ASIA_090422_Running%20on%20Empty_web.pdf*

Kugelman, Michael (2008). "Introduction", in Michael Kugelman and Robert M. Hathaway (eds.), *Hard Sell: Attaining Competitiveness in Pakistani Trade.* Washington D.C.: Woodrow Wilson Centre. *http://www.wilsoncenter.org/topics/pubs/ASIA_Hard.Sell.pdf*

———. (2010). "Water Shortage: The Real Culprit", *Dawn*, July, 26. *http://www.dawn.com/wps/wcm/connect/dawn-content-library/dawn/the-newspaper/editorial/water-shortage-the-real-culprit-670*

Pietz, David (2010). "Managing Scarcity: The Origins and Implications of China's Water Crisis", Paper presented at *Asia Policy Assembly*, the inaugural conference of the National Asia Research Program. Washington D.C. June 17-18.

Polgreen, Lydia and Sabrina Tavernise (2010). "Water Dispute Increases India-Pakistan Tension", *New York Times*, July 20. *http://www.nytimes.com/2010/07/21/world/asia/21kashmir.html*

Salman, Salman M.A. (2008). "The Baglihar Difference and its Resolution Process: A Triumph for the Indus Waters Treaty?", *Water Policy* 10: 115.

Smith, Graeme (2010). "India-Pakistan Water Treaty Poised to Burst", *Globe and Mail*, July 27. *http://www.theglobeandmail.com/news/world/asia-pacific/india-pakistan-water-treaty-poised-to-burst/article1652763/*

———. (2010a). "In Kashmir, Water Treaty Means Less Power to the People", *Globe and Mail*, July 27. *http://www.theglobeandmail.com/news/world/in-kashmir-water-treaty-means-less-power-to-the-people/article1653880/*

Specter, Michael (2006). "The Last Drop: Confronting the Possibility of a Global Catastrophe", *New Yorker*, October 23. *http://www.newyorker.com/archive/2006/10/23/061023fa_fact1*

Strahorn, Eric (2010). "The Indus River Basin in the 21st Century", Paper presented at *Asia Policy Assembly*, the inaugural conference of the National Asia Research Program, Washington DC. June 17-18 June.

Strategic Foresight Group (2010). "Executive Summary", *The Himalayan Challenge: Water Security in Emerging Asia.* Mumbai. *http://www.strategicforesight.com/Himalayan%20Challenge%20ES.pdf*

Times of India (2009). "India's Economic Boom Threatens Water Crisis: Study", *Times of India*, August 13. *http://timesofindia.indiatimes.com/india/Indias-economic-boom-threatens-water-crisis-Study/articleshow/4888373.cms*

United Nations Environment Programme (UNEP) and Asian Institute of Technology (2009). *South Asia: Freshwater Under Threat-Vulnerability Assessment of Freshwater Resources to Environmental Change.* Nairobi, Kenya: UNEP. *http://www.roap.unep.org/pub/southasia_report.pdf*

World Bank (2006). *South Asia: Growth and Regional Integration.* Report #37858-SAS. Washington, D.C.: World Bank/Poverty Reduction and Economic Management Sector Unit, South Asia Region.

13

Interstate Transboundary Water Sharing in India

Conflict and Cooperation

N. SHANTHA MOHAN AND
SAILEN ROUTRAY

Introduction

Water has no respecter of boundaries. Most of the larger rivers in India meander through the administrative boundaries of the Indian federal system. Sometimes, the river itself is the boundary, such as the Indravati which delineates the boundary between Maharashtra and Chhattisgarh for a part of its flow. Some rivers mark metaphorical boundaries as well. River Ganga serves as the vehicle to heaven, whereas River Vaitarani marks the crossing from this world of mortals to the infernal one. Therefore, in a fundamental sense, all rivers are transboundary.

But for are our more somewhat mundane discussion, it's the wayward rivers that do not obey the diktats of human cartographic exercises that end up being marked and categorised as transboundary. For this discussion, the rivers that arise in one province in India but end up in another are termed as such. All of the longer and major rivers in India are transboundary rivers: Mahanadi originates in Amarkantak in Chhattisgarh and crosses over into Odisha before finding its way to the Bay of Bengal; Chambal rises near Mhow in Madhya Pradesh before meandering for more than 900 kilometres to the Yamuna in Uttar Pradesh after having acquired a formidable reputation as the river of the badlands.

Chambal is an example of a long river (with a length of around 960 km) which complicates the ways in which rivers in India are clubbed together and categorised. It arises in the central highlands and drains into the River Yamuna which, in turn, joins the Ganges. Thus, it forms part of the larger Gangetic River system. But it is difficult to locate it within the four-fold categorisation of rivers into Himalayan, peninsular, inland and small coastal ones flowing into the Arabian Sea.

Ganga, Yamuna, Son, Gandak, Brahmaputra, Lohit and Teesta are the major Himalayan rivers. Large parts of the water that these Himalayan rivers receive is from the snowmelt during summer and, therefore, are perennial in nature. Most of the larger rivers in peninsular India are east-flowing, apart from a few exceptions such as the Narmada and Tapti that drain into the Arabian Sea. The important east-flowing rivers of peninsular India are Subarnarekha, Mahanadi, Brahmani, Godavari, Krishna, Cauvery and Penner. The Western Ghats form an important watershed for the southern part of the country and, apart from the several east-flowing rivers that originate here, many small and fast-flowing rivers such as Zuari, Mandovi, Netravati and Periyar originate here and after flowing fast for a short distance drain into the Arabian Sea. Most of the other rivers in India are transboundary. This is true both for larger rivers, such as the Ganga and smaller rivers like Penner. Rivers such as Ghaggar and Luni do not find an outlet into the sea and lose their way in the desert wastes of Rajasthan and Gujarat.

A large number of transboundary rivers have significant implications for water usage and policymaking. This is especially so because India has around 16 per cent of the population and 2.45 per cent of the land area of the world, but has only four per cent of its water resources. In gross national terms, the availability of water can be termed comfortable. But this situation can change with increased demands due to the changing patterns of economic growth and urbanisation. Availability of water can greatly vary in terms of both spatial and temporal aspects. Spatially speaking, the northern and eastern parts of the country are better endowed as compared to the western and southern parts. The less endowed regions are the arid areas in the states of Rajasthan, Gujarat, Maharashtra, Karnataka, Andhra Pradesh and Tamil Nadu. They lie in one rain-shadow region or the other (Iyer, 2003).

India has a monsoonal climate and the average annual rainfall is 1,170 mm. The annual rate varies from less than 150 mm/year in north-western Rajasthan to more than 10,000 mm/year of rainfall in Meghalaya. A large part of the country receives rain for only 100 hours a year. More than half of the precipitation is received in a rainfall of less than about 20 hours (Agarwal, Narain and Sen, 1999). Therefore, the storing of water for use later is of utmost importance. It is this imperative to store water that creates potentials for conflicts over transboundary rivers.

As noted above, India has only 2.45 per cent of land resources and 4 per cent of water resources of the world. India supports 16 per cent of the human population and 15 per cent of the livestock population of the world. India gets around 4,000 bcm of precipitation in an average year. Out of this, 3,000 bcm is received as rainfall in the summer monsoons spread over around 14 weeks. The average annual water potential is around 1,869 bcm, out of which only 690 bcm can be harnessed due to hydrological and geological limitations. Replenishable groundwater resources are estimated at about 433 bcm. Thus, the total utilisable water potential stands at around 1,123 bcm. The utilisation now stands at around 605 bcm out of which irrigation accounts for more than 80 per cent. Domestic, industrial, energy and other sectors consume around 30 bcm, 20 bcm, 20 bcm and 34 bcm respectively. Assured utilisable quantum of water of the country is around 1,123 bcm, whereas estimations of water needs for the year 2050 put this quantum at 1,447 bcm. The total storage capacity is 213 bcm from projects already completed. Storage capacity of around 76 bcm is under construction. A large number of corrective measures need to be taken for appropriate regulation, improvements in efficiencies. Due to spiralling water demand, there is increasing pressure to create storage facilities on rivers, of which most of the larger ones are interstate transboundary rivers (Jeyaseelan, 2010).

Conflict and Cooperation over Transboundary Rivers in India

A large number of rivers in India flow across international and interstate boundaries and many of these are sources of potential conflicts. But the experience surrounding sharing of both international and interstate transboundary river waters is not all grim. The Indus Water Treaty (IWT) between India and Pakistan that came out of a process of mediation, facilitated by the World Bank, is an important example of a working and successful resolution of disputes over an international transboundary river to which India was a party. The treaty awarded nearly 80 per cent of the water of the river system to Pakistan and 20 per cent to India. The treaty has survived three wars between the two countries, showing that such a treaty can weather even the worst of situations in a politically volatile region. The dispute between India and Bangladesh over the Ganga, especially the one surrounding the Farakka Barrage was addressed with the signing of a 30-year water-sharing treaty in 1996. This was an important step towards figuring out mechanisms for sharing the waters of other transboundary rivers between the two countries on a mutually acceptable basis (Mohan, 2010).

Examples of successful dispute resolution of river waters are to be found with regard to interstate transboundary rivers as well, such as the Damodar, Gandak and Subarnarekha. A salient example is the understanding reached on complex issues pertaining to the multi-basin and multipurpose Parambikulam-Aliyar project, where a Joint Water Regulation Board has been established by the riparian states. But it must be mentioned here that, these examples aside, there are many instances of interstate disputes over water sharing that are facing serious situations of potential conflict (Ibid.).

As the preceding discussions have indicated, though India has a long history of cooperation over interstate transboundary rivers, its recent history is fraught with conflicts. There has been bitter fighting over the waters of the Yamuna, Krishna and the Cauvery.

The Yamuna is the longest tributary of the Ganges and is a vital source of water for irrigation and urban use in northern India. It feeds the northern states of Uttar Pradesh, Himachal Pradesh, Haryana, Rajasthan and Delhi. The total present claims on the river are more than twice the total water available. Its waters were shared between Uttar Pradesh and Punjab until 1954, when the new state of Haryana was created by slicing off a part of Punjab. Now Uttar Pradesh controls the Eastern Yamuna Canal and Haryana the Western Yamuna Canal. Following the rising demand from an explosively growing and urbanising state of Delhi, this arrangement was increasingly brought into question, leading to conflicts between Delhi, Haryana and Uttar Pradesh over the sharing of Yamuna water, especially during the lean summer months. Matters have landed up in the courts including the Supreme Court of India through the public interest litigation (PIL) route. With water demand continuing to grow in the basin states, especially in Delhi, there is very little chance of the dispute coming to a closure soon (Swain, 2010).

In peninsular India, the Krishna waters have long been the cause of disputes. It is the second longest river in the south and feeds the states of Maharashtra, Karnataka and Andhra Pradesh. The reorganisation of these states on a linguistic basis in the 1950s questioned the validity of the 25-year agreement on Krishna (1951), that divided the river between the then Bombay, Hyderabad, Mysore and Madras states. The Krishna Dispute Tribunal headed by Justice Bachawat gave its award in 1976, asking the states to utilise their allocations by the year 2000. This led to a race for utilisation of the river's water. One of the repercussions of such a process has been the growing demands and attendant conflicts, exemplified in the conflicts between Andhra Pradesh and Karnataka over the Almatti Dam in the latter state. The cause of the dispute was Karnataka's decision to raise the height of the dam from its original 519 metres to 524.25 metres, a step that would have reduced the capacity of the Nagarjunasagar and Srisailam projects in Andhra Pradesh. But the two states were quick to join hands to oppose Maharashtra, the upper riparian state, when it tried to increase its storage capacity for the waters allocated to it. The concerned states have been complaining against each other to the Central government over the issues. With increasing intensity of resource utilisation, such conflicts can only increase, since the overutilisation of waters of Krishna River Basin is about the highest in peninsular India (Ibid.).

Cauvery is another southern river whose waters have been the cause both of cooperation and conflict over a period of time. The regions of the present day Tamil Nadu were the first movers in using the water of the river. Before the growth of modern dam-building technologies, the upland areas, which today is the state of Karnataka, used very little of its waters. In the latter half of the 19th century, attempts by the then Mysore princely state to construct a dam and use the waters of Cauvery River led to protests from the Madras Presidency of the British Raj. Negotiations between the two followed and a treaty was signed in 1892. This agreement put on record the projects already taken up, and laid down that the Government of Mysore will undertake no new projects. When Mysore proposed the construction of the Krishnaraja Sagar Dam on the Cauvery, the Madras government challenged the decision and sent a complaint to the arbitration committee set up under the agreement of 1892. On receiving an unfavourable judgement from the committee, the Government of Madras took the matter to the Secretary of State in 1919 and got a favourable response. Soon after this, matters were again taken up for negotiation between the two governments and a 50-year agreement was reached in 1924. This agreement allowed the construction of the Krishnaraja Sagar dam in the then Mysore state and the Mettur Dam in Madras Presidency. It also provided a framework for the development of irrigation in the Cauvery Basin. This agreement was supposed to be renewed after 50 years, that is, in 1974, but this was not done. This 50-year period saw the intensification of irrigation development in Karnataka and Tamil Nadu, the successors of the princely state of Mysore and British Indian province of Madras respectively. The increase in the intensive use of water, especially for irrigation, led to conflicts. Tamil Nadu, which had enjoyed the advantage of being the first mover in irrigation development now started complaining about the overuse of the waters of the Cauvery by Karnataka, the upper riparian. It started demanding the creation of a tribunal for resolution of these disputes and sharing of the waters of Cauvery. In 1990, the Cauvery Water Disputes Tribunal was set up, which came out with its award in 2007. The Government of Karnataka was dissatisfied with the award (Settar, 2010).

Rules and Mechanisms to Address Transboundary Water Sharing in India

The above section shows that the history of interstate river water sharing has been characterised by both cooperation and conflict. Water conflicts are of many types, depending upon the perspectives of the contesting parties and the nature of the contest. The complexity of the issues is exacerbated by the deficiencies in the legal and institutional mechanisms. Irrigation as a sector consumes more than 80 per cent of all available water in the country. Listed as Enry 17 under the State List in the Constitution, it is subject to the provisions of entry 56 of the Union List. The latter empowers the Central government to legislate on interstate river issues. But this entry has been underused. Article 262 of the Constitution provides for an adjudication role by the Centre in these conflicts. The Inter-State Water Disputes Act (ISWD), 1956, was promulgated under Article

262. This Act provides for the formation of tribunals for settling transboundary river disputes (Mohan, 2010).

According to the provisions of the ISWD Act, a state government having a dispute with another over interstate river waters can ask the Centre to se up a tribunal for adjudication. The tribunal is to comprise a chairperson and two members. These three persons, to be nominated by the Chief Justice of India, must be judges of that august court at the time of nomination. The tribunal is empowered to appoint assessors whose task would be to assist in the investigation and to advise the tribunal. The Act mandates the publication of the tribunal's award. Its decision is final, and is binding on the parties to the dispute. The tribunals set up for settling the disputes surrounding the Krishna, Godavari and Narmada rivers are perceived as having been relatively successful. But the efficacy of tribunals to settle such disputed rivers is coming increasingly under question. There have been substantial problems surrounding the tribunals set up to settle the disputes over the waters of Ravi-Beas and Cauvery. The awards in both the cases have failed to resolve the disputes and have led to further bouts of intense politicking. The tribunal's award now has the status of the decree of the Supreme Court, by virtue of the recent amendments to the ISWD Act. The tribunals take a lot of time to reach a final settlement. Even the amended Act of 2002 mandates a time limit of six years. But even six years is, relatively speaking, a long period of time to resolve such issues. In this context, a mention must be made of several non-official civil society efforts that have been working assiduously to address the issues concerning river water sharing. The Madras Institute of Development Studies (MIDS), Chennai, initiated the setting up of a platform to facilitate dialoguing between the farmers of Karnataka and Tamil Nadu in the Cauvery Basin. Thus, by talking to each other, the farmers have started to understand each others' problems and needs (Ibid.).

Addressing Transboundary Water Conflicts

In this context, we list some ways that can be of some help in addressing issues of transboundary water conflicts. The first path is of an institutional nature. We suggest that the already existing institutions, as also the forming of new institutions, such as the River Basin Organisation (RBO), can go some distance in resolving water conflicts. We also need to use some new tools or use old tools differently to be able to deal with water conflicts creatively. In this regard, we list mediation and an alternative approach to scenario building as two possible ways.

The Interstate Council

Article 263 of the Constitution envisages the establishment of an Inter-State Council (ISC) with the mandate to enquire into and give advice on disputes arising between states, to investigate subjects of common interest to the states and to offer recommendations for better coordination of policy and action among the states concerned. The Administrative Reforms Commission (1969), Rajamannar Committee (1971) and the Sarkaria Commission (1983) in their reports recommended the setting up of the ISC. The Inter-State Council was finally established by a Presidential Order on May 28, 1990, as a recommendatory body to fulfill the already mentioned Constitutional mandate. The council comprises the Prime Minister, all chief ministers, administrators of Union territories, six ministers of the Central cabinet, as also permanent invitees. The Inter-State Council Secretariat has been empowered to deal with any matter in the Union List, Concurrent List or the State List of the Constitution where there exists a common interest. The ISC Secretariat, which provides organisational support to the council, is headed by a secretary who is also the secretary of the Commission on Centre-State Relations. He is assisted by two advisers and additional secretaries.

Conducting relevant studies is also a mandatory task of the ISC. Studies have already been commissioned on: providing compensation to resource bearing states; subnational governance and an appraisal of measures taken to implement the Directive Principles of State Policy. The council provides a forum for discussions on complex public policy and governance issues having a bearing on Centre-state relations or having Interstate dimensions. Because the council is a statutory body, and has now built a body of experience in dealing with matters that are of common interest to states, it can play a useful role in facilitating dialogues and discussions towards resolving conflicts surrounding transboundary water sharing in the country (Iyer, 2002; Mohan, 2010; Interstate Council, 2011).

River Basin Organisations

There is a need to look at arbitration and negotiations as methods of conflict resolution. One institutional arrangement that can be used to facilitate negotiation surrounding interstate rivers is the River Basin Organisation (RBO). RBOs can be set up under the River Boards Act (RBA) of 1956, which was legislated under Article 56 of the Union List. These are empowered to regulate and develop interstate rivers and their basins. The board is required to have members with expertise in irrigation, water-soil conservation and finance. But river boards have not been established because the state governments apprehend that the boards would impinge upon their authority and power (Iyer, 2007). However, in this era of coalition politics, the states have started to see the advantage as also the need to set up RBOs.

Mediation as a Method for Dealing with Transboundary Water Conflicts

To date, seven tribunals have been established to deal with disputes over the waters of transboundary rivers. But these have not always helped resolve the disputes in a satisfactory manner. These tribunals depend upon the legal principles of arbitration. Their awards, although supposedly final and binding, have been challenged in courts and the courts have entertained these challenges. The judicial process is essentially an adversarial process and damages the relationship between the disputants. In contrast, mediation is a process that employs a neutral person or persons to facilitate a process of negotiations between the disputing parties so as to arrive at a mutually acceptable solution. Mediators should not have any direct interest in the conflict, or control over the process of mediation, or its outcomes. The power is vested in the disputants, but mediation process is flexible, informal and can draw upon the multidisciplinary perspectives of the mediators. At the subcontinental level, the World Bank played the role of mediator between India and Pakistan, and succeeded in resolving the conflicts surrounding the rivers of the Indus Basin. In the Zambezi River dispute, the Vatican negotiated to fruition an 11-nation agreement for the joint use and management of the river's waters (Devi, 2010).

Given the above examples, there is an urgent need in India to deploy mediation as a tool for conflict resolution and participatory management.

Alternative Approaches to Scenario-Building for Water

The manner in which scenario-building in the water sector takes place has reduced it to a mere 'technical' tool for prediction. Scenario-building is not a tool for projection, and need not be used as one. A scenario is essentially an imaginative exercise involving political and social choices. It is as much a tool for action as it is for thought. While undertaking exercises of scenario-building, one needs to take into account the physical qualities of water as a resource. Generally, in such exercises, the current patterns of consumption are taken for granted to arrive at projections of various likely demands in the future. Thus, these attempts are just tunnel vision exercises based on the current patterns to project possible future demands. We argue that there is a need for a completely different way of building scenarios. We need to hypothetically freeze the total available water or freeze the quantum at current levels of total consumption in a given region—the unit for analysis—and build scenarios of alternative usage patterns. Thus, building a scenario should not be taken as merely predicting the total quantum of water that would be in demand at a future date. What one is trying to do is to plan as if water, and its characteristics as a life-giving resource, mattered (which in point of fact it does). This will necessarily be a non-technocratic and democratic exercise, since what one is trying to simulate in this proposed alternative method is the social choices that we might want to make if water availability and/or consumption were to be frozen at some arbitrary point in the present. Such an exercise will help us in unravelling the assumptions that we make while making projections. Such an approach will also help us radically interrogate theories of risk society by positing scenarios as 'designs' (Routray, 2010).

Conclusions

Water is becoming an increasingly important site of contestation among the states in India because of the rapid rates of population and economic growth and because of increasing urbanisation. The growing importance of coalitions at the national level and the related assertion of regional identities add to the intractability of the problems surrounding transboundary water sharing in India. More often than not, such issues are a result of the focus on demand-side management. Many scholars have argued that emphasising on supply-side management might be one way of dealing with such issues. There is a lot of merit in this argument. But we need to undertake institutional innovations as well. The suggestions for setting up the RBOs and giving greater role to the Interstate Council in dealing with interstate river issues should be given due consideration. With the changing political dynamics in the country, it should not be difficult to convince the states that the relationship between governments at the level of the states and at the Centre is not a zero-sum game. Increasing roles for Central institutions in dealing with issues emerging out of sharing the waters of the transboundary rivers does not necessarily mean a whittling down of the powers of the states. Secondly, as this paper has argued, one needs to creatively use already existing tools (such as mediation and scenario-building exercises) for managing water resources of interstate rivers more effectively and democratically. The emergent solutions have to be context-specific and have to borrow creatively from a bouquet of solutions.

References

Agarwal, Anil, Sunita Narain and Srabani Sen (eds.) (1999). *The State of India's Environment: The Citizen's Fifth Report*. New Delhi: Centre for Science and Environment.

Devi, Geetha M. (2010). "Legal Framework for Resolution of Water Disputes", Paper presented at *The National Consultation on Water Conflicts in India: The State, the People and the Future*. March 15-16. Bangalore: NIAS.

Interstate Council (2011). *http://interstatecouncil.nic.in/* (last accessed on 01.09.2011, 19.46 hours)

Iyer, Ramaswamy (2007). *Towards Water Wisdom: Limits, Justice, Harmony*. New Delhi: Sage Publications.

———. (2003). *Water: Perspectives, Issues, Concerns*. New Delhi: Thousand Oaks and London: Sage Publications.

———. (2002). "Inter-State Water Disputes Act 1956 Difficulties and Solutions", *Economic and Political Weekly* 37(28): 2907-2910.

Jeyaseelan, R. (2010). "Regulatory Aspects in Water Resources Development and Management", in Shantha Mohan, Sailen Routray and N. Shashikumar (eds.), *River Water Sharing: Transboundary Conflict and Cooperation in India*. New Delhi: Routledge. pp.81-95.

Mohan, Shantha N. (2010). "Locating Transboundary Water Sharing in India", in N. Shantha Mohan, Sailen Routray and N. Shashikumar (eds.), *River Water Sharing: Transboundary Conflict and Cooperation in India*. New Delhi: Routledge. pp.3-22.